DATE DUE

BRODART, CO.   Cat. No. 23-221

In Pursuit of the Traveling Salesman

In Pursuit of the Traveling Salesman

Mathematics at the Limits of Computation   William J. Cook

PRINCETON UNIVERSITY PRESS
Princeton and Oxford

Copyright © 2012 by Princeton University Press

Published by Princeton University Press, 41 William Street,
Princeton, New Jersey 08540
In the United Kingdom: Princeton University Press, 6 Oxford Street,
Woodstock, Oxfordshire OX20 1TW
press.princeton.edu

Library of Congress Cataloging-in-Publication Data
Cook, William, 1957–
In pursuit of the traveling salesman : mathematics at the limits of computation / William J. Cook.
    p. cm.
Includes bibliographical references and index.
ISBN 978-0-691-15270-7 (hardback)
1. Traveling salesman problem.   2. Computational complexity.   I. Title.
QA164.C69 2012
511'.5—dc23   2011030626

British Library Cataloging-in-Publication Data is available

This book has been composed in Minion
Printed on acid-free paper ∞

Typeset by S R Nova Pvt Ltd, Bangalore, India
Printed in the United States of America

10 9 8 7 6 5 4 3 2 1

Listen, mate, I've traveled every road in this here land.

—Geoff Mack, Lyrics to "I've Been Everywhere."

Contents

Preface

The star of Geoff Mack's song has been to Reno, Chicago, Fargo, Buffalo, Toronto, Winslow, Sarasota, Wichita, Tulsa, Ottawa, Oklahoma, Tampa, Panama, Mattawa, LaPaloma, Bangor, Baltimore, Salvador, Amarillo, Tocapillo, Barranquilla, and Padilla.

One night in February 1988, my friend Vašek Chvátal and I decided to follow in the footsteps of mathematical giants and take a crack at finding the shortest way around such destinations. Next day we met at Tri-State Camera, a computer vendor in lower Manhattan. When the technician learned we were mathematicians in need of a speedy computer, he looked us in the eye and warned, "You guys aren't trying to solve that traveling salesman problem, are ya?" Quite a bit of foresight there. This was the first of many computers we ground to a halt, spending the better part of the next twenty years searching for solutions.

The notorious problem is to compute the shortest route to visit each city on a specified list and return to the starting point. In the case of Geoff Mack's traveler, one could conceivably check all 51,090,942,171,709,440,000 tours through the twenty-two cities. This computation would require even the fastest of supercomputers to roll up its sleeves and prepare for a hard day's work, but with patience it may be possible to carry out the search. If, however, we had one hundred cities or so, then checking all routes to select the shortest is out of the question, even devoting the entire planet's computing resources to the task.

This observation on sorting through all possible solutions is by no means a convincing argument that the salesman problem is actually difficult. There are similar problems that are very easy to solve and yet have far more candidate solutions than the number of ways to route a salesman. What sets the traveling salesman problem apart is the fact that despite decades of research by top applied mathematicians around the world, in general it is not known how to significantly improve upon simple brute-force checking. It is a real possibility that there may never exist an efficient method that is guaranteed to solve every example of the problem. This is a deep mathematical question: is there an efficient solution method or not? The topic goes to the core of complexity theory concerning the limits of feasible computation. For the stouthearted who would like to tackle the general version of the problem, the Clay Mathematics Institute will hand over a $1,000,000 prize to anyone who can either produce an efficient method or prove that this is impossible.

The complexity question that is the subject of the Clay Prize is the Holy Grail of traveling-salesman-problem research and we may be far from seeing its resolution. But this is not to say mathematicians have thus far come away empty-handed. Indeed, the problem has led to a large number of results and conjectures that are both beautiful and deep. In the arena of exact computation, an 85,900-city challenge problem was solved in 2006, when the optimal tour was pulled out of a mind-boggling number of candidates in a computation that took the equivalent of 136 years on top-of-the-line computer workstations. On the practical side, solution methods are used to compute optimal or near-optimal tours for a host of applied models on a daily basis.

One of the enduring strengths of the traveling salesman problem has been its remarkable success as an engine of discovery in applied mathematics and its fields of operations research and mathematical programming. And many more discoveries could be waiting just around the corner. A primary goal of the book is to stimulate readers to pursue their own ideas for the solution of this mathematical challenge.

In setting down these notes I have had the pleasure of receiving help and support from many people. First of all, I thank my comrades David Applegate, Robert Bixby, and Vašek Chvátal, for over twenty years of fun and work toward unraveling part of the traveling salesman mystery. I also thank Michel Balinsky, Mark Baruch, Robert Bland, Sylvia Boyd, William Cunningham, Michel Goemans, Timothy Gowers, Nick Harvey, Keld Helsgaun, Alan Hoffman, David Johnson, Richard Karp, Mitchel Keller, Anton Kleywegt, Bernhard Korte, Harold Kuhn, Jan Karel Lenstra, George Nemhauser, Gary Parker, William Pulleyblank, Andre Rohe, Lex Schrijver, Bruce Shepherd, Stan Wagon, David Shmoys, Gerhard Woeginger, and Phil Wolfe for discussions of the problem and its history.

Images and historical material used in the presentation were provided by Hernan Abeledo, Leonard Adleman, David Applegate, Masashi Aono, Jessie Brainerd, Robert Bixby, Adrian Bondy, Robert Bosch, John Bartholdi, Nicos Christofides, Sharlee Climer, James Dalgety, Todd Eckdahl, Daniel Espinoza, Greg Fasshauer, Lisa Fleischer, Philip Galanter, Brett Gibson, Marcos Goycoolea, Martin Grötschel, Merle Fulkerson Guthrie, Nick Harvey, Keld Helsgaun, Olaf Holland, Thomas Isrealsen, David Johnson,

Michael Jünger, Brian Kernighan, Bärbel Klaaßen, Bernhard Korte, Drew Krause, Harold Kuhn, Pamela Walker Laird, Ailsa Land, Julian Lethbridge, Adam Letchford, Panagiotis Miliotis, J. Eric Morales, Randall Munroe, Yuichi Nagata, Denis Naddef, Jaroslav Nešetřil, Manfred Padberg, Elias Pampalk, Rochelle Pluth, Ina Prinz, William Pulleyblank, Gerhard Reinelt, Giovanni Rinaldi, Ron Schreck, Éva Tardos, Mukund Thapa, Michael Trick, Marc Uetz, Yushi Uno, Günter Wallner, Jan Wiener, and Uwe Zimmermann. I thank them all for their kindness.

This book was written in the great environments of the H. Milton Stewart School of Industrial and Systems Engineering at Georgia Tech and the Department of Operations Research and Financial Engineering at Princeton University. My work on the traveling salesman problem is funded by grants from the National Science Foundation (CMMI-0726370) and the Office of Naval Research (N00014-09-1-0048), and by a generous endowment from A. Russel Chandler III. I am grateful for their continued support.

Finally, I thank my family, Monika, Benny, and Linda, for years of patiently listening to salesman stories.

In Pursuit of the Traveling Salesman

1: Challenges

*It grew out of the trio's efforts to find solutions for a classic
mathematical problem—the "Traveling Salesman"
problem—which has long defied solution by man, or by the fastest
computers he uses.*
—IBM Press Release, 1964.[1]

An advertising campaign by Procter & Gamble caused a stir among applied mathematicians in the spring of 1962. The campaign featured a contest with a $10,000 prize. Enough to purchase a house at the time. From the official rules:

> Imagine that Toody and Muldoon want to drive around the country
> and visit each of the 33 locations represented by dots on the contest
> map, and that in doing so, they want to travel the shortest possible
> route. You should plan a route for them from location to location
> which will result in the shortest total mileage from Chicago, Illinois
> back to Chicago, Illinois.

Police officers Toody and Muldoon navigated *Car 54* in a popular American television series. Their 33-city task is an instance of the *traveling salesman problem*, or *TSP* for short. In its general form, we are given a collection of cities and the distance to travel between each pair of them. The problem is to find the shortest route to visit each city and to return to the starting point.

Is the general problem easy, hard, or impossible? The short answer is that no one really knows. This is both the mystery and attraction of this famous challenge in computational mathematics. And much more than a struggling salesman is at stake. The TSP is the focal point of a larger debate on the nature of complexity and possible limits to human knowledge. If you are ready for action, then a sharp pencil and a clean piece of paper are all you may need to give a helping hand to the salesman and possibly to make a quantum leap in our understanding of the world in which he or she travels.

Figure 1.1
Car 54 contest. Image
courtesy of Procter &
Gamble.

Tour of the United States

Despite its nasty reputation, the TSP is an easy enough task from one perspective: there are only finitely many possible routes through a given set of cities. So a 1962-era mathematician could have checked each possible Toody-Muldoon tour, recorded the shortest, sent the solution to Procter & Gamble, and waited for the $10,000 check to arrive in the mail. A simple and flawless strategy. With one possible catch. The number of distinct tours is exceedingly large to consider checking one by one.

This difficulty was noticed in 1930 by the Austrian mathematician and economist Karl Menger, who first brought the challenge of the TSP to the attention of the mathematics community. "This problem is of course solvable by finitely many trials. Rules that give a number of trials below the number of permutations of the given points are not known."[2] A tour can be specified by announcing the order in which the cities are to be visited. For example, if we label the 33 destinations of Toody and Muldoon as A through Z and 1 though 7, that is, A for Chicago, B for Wichita, etc., then we can record a possible tour as

ABCDEFGHIJKLMNOPQRSTUVWXYZ1234567

or any other arrangement of the 33 symbols. Each such arrangement is a *permutation* of the symbols. The ordering implied by the arrangement is circular, in that we travel from the last city back to the first. So we can record the same tour in 33 ways, depending on which city we put in the first position. To avoid such overcounting, we may as well always start with city *A*. This leaves 32 choices for the second city, 31 choices for the third city, and so on. Altogether, we have $32 \times 31 \times 30 \times \cdots \times 3 \times 2 \times 1$ tours to consider. This is the total number of permutations of 32 objects. It is written as 32! and spoken as 32 *factorial*.

In the Procter & Gamble contest we can save effort by noting that the distance to travel between Chicago and Wichita is the same as the distance between Wichita and Chicago, and this is true also for every other pair of cities. With such symmetry it does not matter in which direction we travel around a tour, so an ordering

ABCDEFGHIJKLMNOPQRSTUVWXYZ1234567

is the same as its reverse

7654321ZYXWVUTSRQPONMLKJIHGFEDCBA.

We can therefore cut down by half our count of the 33-city tours, leaving only 32!/2 orderings to check. Before you go ahead and get out your Ticonderoga #2 pencil, note that this is

131,565,418,466,846,765,083,609,006,080,000,000

distinct tours that we must examine.

These days we would of course employ a computer to run through the list. So let's choose a big one, the $133,000,000 IBM Roadrunner Cluster of the United States Department of Energy. This 129,600-core machine topped the 2009 ranking of the 500 world's fastest supercomputers, delivering up to 1,457 trillion arithmetic operations per second.[3] Let's assume we can arrange the search for tours such that examining each new one requires only a single arithmetic operation. We would then need roughly 28 trillion years to solve the 33-city TSP on the Roadrunner, an uncomfortable amount of time, given that the universe is estimated to be only 14 billion years old. No wonder Menger was unsatisfied with the brute-force solution to the problem.

When considering the implications of this quick analysis, we must keep in mind that Menger writes only that faster rules for solving the

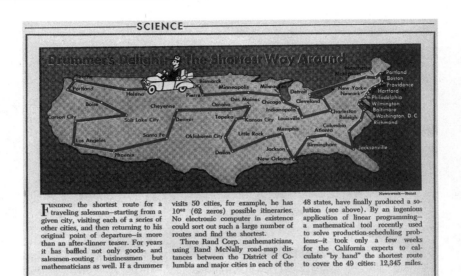

—SCIENCE—

FINDING the shortest route for a traveling salesman—starting from a given city, visiting each of a series of other cities, and then returning to his original point of departure—is more than an after-dinner teaser. For years it has baffled not only goods- and salesmen-routing businessmen but mathematicians as well. If a drummer visits 50 cities, for example, he has 10^{62} (62 zeros) possible itineraries. No electronic computer in existence could sort out such a large number of routes and find the shortest.

Three Rand Corp. mathematicians, using Rand McNally road-map distances between the District of Columbia and major cities in each of the 48 states, have finally produced a solution (see above). By an ingenious application of linear programming—a mathematical tool recently used to solve production-scheduling problems—it took only a few weeks for the California experts to calculate "by hand" the shortest route to cover the 49 cities: 12,345 miles.

Newsweek—Bendt

Figure 1.2
Drummer's Delight. *Newsweek,*
July 26, 1954, page 74.

salesman problem are unknown, not that such rules are out of the question. John Little and coauthors sum this up nicely in the following comment on the Procter & Gamble contest. "A number of people, perhaps a little over-educated, wrote the company that the problem was impossible—an interesting misinterpretation of the state of the art."[4] Little et al. went on to describe a breakthrough in TSP solution methods, but they could not push their computer codes far enough to actually solve the 33-city challenge. It appears that no one in the country was able to produce a route that could be guaranteed to be the shortest of all possible tours for Toody and Muldoon.

The 33-city problem was definitely a tough nut to crack, but if we turn back the clock to 1954, then we find a team that almost certainly would be able to deliver the optimal route, together with a written guarantee that their solution is the shortest. The team tackled a larger touring problem through the United States, visiting a city in each of the 48 states, as well as Washington, D.C. This particular challenge had been circulating through the mathematics community since the mid-1930s. Its solution was reported in *Newsweek*.[5]

Finding the shortest route for a traveling salesman—starting from a given city, visiting each of a series of other cities, and then returning to his original point of departure—is more than an after-dinner teaser. For years it has baffled not only goods- and salesman-routing businessmen but mathematicians as well. If a drummer visits 50 cities, for example, he has 10^{62} (62 zeros) possible itineraries. No electronic computer in existence could sort out such a large number of routes and find the shortest.

Three Rand Corp. mathematicians, using Rand McNally road-map distances between the District of Columbia and major cities in each of the 48 states, have finally produced a solution. By an ingenious application of linear programming—a mathematical tool recently used to solve production-scheduling problems—it took only a few weeks for the California experts to calculate "by hand" the shortest route to cover the 49 cities: 12,345 miles.

The California experts were George Dantzig, Ray Fulkerson, and Selmer Johnson, part of an exceptionally strong and influential center for the new field of mathematical programming, housed at the RAND Corporation in Santa Monica.

The RAND team's guarantee involves some pretty mathematics that we take up later in the book. For now it is best to think of the guarantee as a proof, like those we learned in geometry class. The Dantzig et al. proof establishes that no tour through the 49 cities can have length less than 12,345 miles. Matching the proof with their tour of precisely this length shows that this particular instance of the TSP has been settled, once and for all.

Dantzig and company missed out on the $10,000 contest, but we can report that a computer implementation of their ideas makes easy work of the 33-city TSP. A shortest route for Toody and Muldoon is depicted in Figure 1.3. Although no one in 1962 knew for certain that this was the shortest possible tour, a number of contestants did find and report this

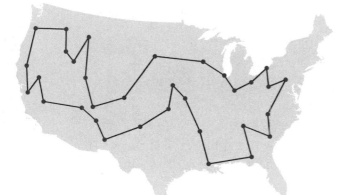

Figure 1.3
Optimal tour for
Car 54 contest.

same ordering. Among the people tied for first place in the contest were mathematicians Robert Karg and Gerald Thompson, who created a hit-or-miss heuristic strategy that produced the winning solution.[6] And the story has a happy ending, at least for the mathematics community. As a tiebreaker, contestants were asked to write a short essay on the virtues of one of Procter & Gamble's products. Thompson's prose on soaps took a grand prize.

An Impossible Task?

The RAND team's work put an end to the 48-states challenge, but it did not finish off the TSP. One big success did not imply the team could handle other, possibly larger, instances of the problem. In fact, if Las Vegas were taking bets on the outcome, the odds-on favorite among mathematicians would be that we will never fully solve the TSP. We must be careful here. By a solution we mean an *algorithm*, that is, a step-by-step recipe for producing an optimal tour for any example we may throw at it. Just finding the best route through the United States or any other country does not do the job.

Picking up on the expected difficulty of the general TSP challenge, the science-fiction story "Antibodies", by Charles Stross, chronicles doomsday events following the discovery of an efficient solution method for the salesman.[7] One can hope that a brilliant insight into the TSP will not signal the end of the world as we know it, but it will certainly turn the planet upside down and give it a good shake. To see why, let's start with a series of quotes.

> 'It seems very likely that quite a different approach from any yet used may be required for successful treatment of the problem. In fact, there may well be no general method for treating the problem and impossibility results would also be valuable.'
>
> —Merrill.Flood, 1956.[8]

> 'I conjecture that there is no good algorithm for the traveling salesman problem.'
>
> —Jack Edmonds, 1967.[9]

> 'In this paper we give theorems which strongly suggest, but do not imply, that these problems, as well as many others, will remain intractable perpetually.'
>
> —Richard Karp, 1972.[10]

The authors of these remarks are three giants of traveling-salesman research. Merrill Flood rallied support for the problem in the 1940s; more than anyone else, Flood is responsible for the emergence of the TSP as a fundamental topic of study. Discussing the state of the problem in 1956, Flood first raised the possibility that efficient methods may simply never exist. This point was hammered home by Jack Edmonds a decade later in what amounts to a mathematical bet against the hope for a general solution method. Edmonds was modest in describing the support for his bet: "My reasons are the same as for any mathematical conjecture: (1) It is a legitimate mathematical possibility, and (2) I do not know." But he is teasing us with these words: Edmonds is one of the profound thinkers in twentieth-century mathematics and he certainly had something deep in mind when placing money against the TSP. Five years later, the true nature of the bet was made clear in a publication by the great computer scientist Richard Karp, connecting the TSP with a host of other computational problems. We save the details of Karp's theory for chapter 9, but a quick account will be enough to understand why the characters of "Antibodies" shuddered at the announcement of a fast TSP algorithm.

Good and Bad Algorithms

When Edmonds writes "good algorithm," he uses the word good in the same way as you and I: an algorithm is good if it can solve problems in an amount of time we find acceptable. For this to make sense in mathematics, however, he had to make "good" into a formal notion. Clearly, we cannot expect every example of the TSP to be solved, say, in under a minute by a human or by one of our machines. We must at least be willing to allow for the solution time to grow as the number of cities grows. The point to be decided is what rate of growth is acceptable.[11]

Figure 1.4
Jack Edmonds, 2009.
Photograph courtesy
of Marc Uetz.

Table 1.1
Running time on a 10^9-operations-per-second computer.

	$n = 10$	$n = 25$	$n = 50$	$n = 100$
n^3	0.000001 seconds	0.00002 seconds	0.0001 seconds	0.001 seconds
2^n	0.000001 seconds	0.03 seconds	13 days	40 trillion years

Let's use the symbol n to indicate the size of a problem; for the TSP this is the number of cities. Reading a list of locations to visit takes time proportional to n, so a tough manager might demand that we produce an optimal tour also in time proportional to n. Such a manager would be wildly optimistic. Edmonds himself allows for faster rates of growth in the running time, but with an insightful break between good and bad. A *good algorithm* is one that comes with a guarantee to complete its work in time at most proportional to n^k for some power k. The power k can be any value, such as 2, 3, or more, but it must be a fixed number—it cannot increase as n gets larger. Thus, a growth rate of n^3 is good, but growth rates of n^n and 2^n are bad. To give you a feeling for this, in table 1.1 we have calculated the running times for a few values of n, assuming a computer can handle 10^9 instructions per second. If $n = 10$, the bad algorithm is fine. But you don't want to be stuck behind a 2^n algorithm if n gets up to 100 or so.

Edmonds's formal notion of "good" might not always agree with our intuition. An algorithm that requires n^{1000} steps is not appealing if we need to solve an instance of the TSP with 100 cities. Nonetheless, his idea revolutionized the study of computing. The precise good/bad dichotomy creates real targets for mathematicians, fueling great interest in computational issues. And on the practical side, once a problem is shown to have a good algorithm, researchers pull out all stops in a race to decrease the value of the power k, typically getting down to running-time bounds proportional to n^2, n^3, or n^4, and computer codes capable of handling large instances.

Figure 1.5
Travelling Salesman
Problem. Image
courtesy of Randall
Munroe, xkcd.com.

Unfortunately for TSP fans, no good algorithm is known for the problem. The best result thus far is a solution method, discovered in 1962, that runs in time proportional to $n^2 2^n$. Although not good, this growth rate is much smaller than the total number of tours through n points, which we know is $(n - 1)!/2$, perhaps satisfying the curiosity of Menger.

The Complexity Classes $\mathcal{P}$ and $\mathcal{NP}$

Edmonds's dichotomy carries over to computational problems, dividing them into those for which good algorithms exist and those for which they do not. The former problems are the ones we like, and they are known collectively as the class $\mathcal{P}$.

Why $\mathcal{P}$ and not $\mathcal{G}$? Well, researchers were not entirely comfortable with the emotional charge that comes with the word "good," and it became standard to use the term *polynomial-time algorithm*. So $\mathcal{P}$ for polynomial.

The definition of $\mathcal{P}$ is clear-cut, but it can be tricky to tell whether or not a given problem belongs to this "good" class. It may well be that the TSP is in $\mathcal{P}$ and we just haven't yet discovered the good algorithm to prove its membership. A glimmer of hope is that at least we know a good tour when we see one. Indeed, suppose our challenge is to find a tour, say, of length less than 100 miles. If someone hands us such a solution, then we can check easily that it does indeed beat the 100-mile target. This property makes the TSP a member of the class known as $\mathcal{NP}$, consisting of all problems for which we can check the correctness of a solution in polynomial time. The pair of letters stands for *non-deterministic polynomial*. The unusual name aside, this is a natural class of problems: when we make a computational request, we ought to be able to check that the result meets our specifications.

The Big Question

Could it be that $\mathcal{P}$ and $\mathcal{NP}$ are two names for the same class of problems? It is possible. An approach for proving this was laid out in a breakthrough result by Stephen Cook in 1971. (No relation to me, although I have enjoyed a number of free dinners due to mistaken identity.) Cook's Theorem states that there exists a problem in $\mathcal{NP}$ such that if we have a good algorithm for this single problem, then there is a good algorithm for every problem in $\mathcal{NP}$. In fact, Cook, Karp, and others have shown that there are many such $\mathcal{NP}$-complete problems, the most prominent being the TSP itself.

Finding a good algorithm for an $\mathcal{NP}$-complete problem would show that $\mathcal{P}$ is equal to $\mathcal{NP}$. Thus, the first person to discover a general method for the TSP will bring home considerably more cash than the winner of the Procter & Gamble contest: the Clay Mathematics Institute has offered a $1,000,000 prize for either a proof or disproof that $\mathcal{P} = \mathcal{NP}$.

The betting line is that the two problem classes are not equal, but there is no great theoretical reason for thinking this is the case. It is simply a feeling that equality is too much to ask: any problem we can formulate in a verifiable manner would immediately have an efficient method of solution. In fact, current encryption systems make use of the assumption that certain $\mathcal{NP}$ problems are difficult to solve. Internet commerce would grind to a halt if there were quick algorithms for these members of $\mathcal{NP}$; this would be like handing code breakers and hackers a Swiss Army knife for snooping data.

The downfall of society in "Antibodies" was more insidious, however, than simply failures in encryption—artificial intelligence programs suddenly became greatly more effective and took over their biological masters. It seems probable we could deal with such pesky machines, and it is likely the good consequences of $\mathcal{P} = \mathcal{NP}$ would greatly outweigh the bad. In a 2009 survey article, Lance Fortnow wrote: "Many focus on the negative, that if $\mathcal{P} = \mathcal{NP}$ then public-key cryptography becomes impossible. True, but what we will gain from $\mathcal{P} = \mathcal{NP}$ will make the whole Internet look like a footnote in history."[12] His argument is that optimization becomes easy, thus salesmen can find their shortest routes, factories can run at peak capacity, airlines can manage their schedules without delays, and so on. Simply put, we will better utilize the resources available in our world. Vastly more powerful tools would also be available in science, economics, and engineering, providing a steady flow of breakthroughs to keep Nobel Prize committees busy for years to come. A rosy world, but the bets are against it.

The resolution of $\mathcal{P}$ versus $\mathcal{NP}$ is clearly one of the great challenges of our time. In approaching an $\mathcal{NP}$-complete problem like the TSP, however, it is important not to get too caught up in possible ramifications of a good solution method. The lofty implications aside, the problem comes down to a simple routing of a salesman. An ingenious idea could turn the scales.

One Problem at a Time

Until someone steps forward with a possibly earth-shattering result on the general complexity question, what is to be done with the TSP? Well, facing the salesman head on, the clear target is the solution of larger and more difficult instances of the problem.

The TSP is the standard bearer of a pragmatic school of research known as *algorithm engineering*.[13] The motto here is to not take no for an answer. Theoretical considerations may suggest that once we reach a certain size there exist instances of the TSP that necessarily take an exorbitant amount of computation, but this does not imply that whenever we see a specific large example we must give up and resort to a rough guess for a tour. Indeed, this take-no-prisoners attitude has led the community to techniques and computer codes capable of solving examples of almost unbelievable complexity.

Knocking off a previously unsolved challenge instance is a heralded event among researchers, akin to scaling a new Himalayan peak or running the 100-meter dash in record time. It is not that we have a desperate thirst for the details of particular optimal tours, but rather a desperate need to know that the TSP can be pushed back just a bit further. The salesman may defeat us in the end, but not without a good fight.

From 49 to 85,900

The heroes of the field are Dantzig, Fulkerson, and Johnson. Despite the dawning of the computer age and a steady onslaught of new researchers tackling the TSP, the 49-city example that Dantzig et al. solved by hand stood as an unapproachable record for seventeen years. Algorithms were developed, computer codes written, and research reports published, but year after year their record held its ground. The long run was finally snapped in 1971 by IBM researchers Michael Held and Richard Karp; the same Karp who studied TSP impossibility results, clearly not satisfied with theory alone. The test instance in this case consisted of 64 points dropped at random into a square region, with travel costs set to the straight-line distances between pairs of points.

The algorithm of Held and Karp reigned supreme for several years, with a number of teams tweaking the method in attempts to squeeze out greater performance. But the Dantzig et al. approach struck back in 1975, when Panagiotis Miliotis eclipsed the Held-Karp record by employing a variant of the original RAND idea to compute the shortest route through 80 random points.

The Miliotis work hinted at the fact that the Dantzig et al. approach might offer possibilities to push well beyond the expected limits of TSP computation. This was reinforced shortly thereafter by theoretical studies by Martin Grötschel and Manfred Padberg, who laid foundations for a great expansion of the basic methodology. This pair of mathematicians went

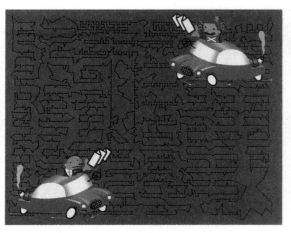

Figure 1.6
A new TSP record, 3,038 cities.
Discover, January 1993.

on to dominate the TSP scene for the next fifteen years. Their successes began with Grötschel's construction of an optimal 120-city tour through Germany, published in his 1977 doctoral thesis. Padberg then teamed up with IBM researcher Harlan Crowder, computing the optimal solution for a 318-city example that arose in a circuit-board drilling application. These two results, although great in their own right, turned out to be only preliminary steps toward a series of startling announcements in 1987, a banner year for the TSP. Working independently on opposite sides of the Atlantic, Grötschel and Padberg led teams that solved in rapid succession instances consisting of 532 cities in the United States, 666 locations in the world, and 1,002-city and 2,392-city drilling problems; Grötschel worked with doctoral student Olaf Holland at the University of Bonn, and Padberg worked with Italian mathematician Giovanni Rinaldi at New York University.

Riding this wave of excitement, Vašek Chvátal and I decided to join the TSP-computation race in early 1988. We were in the unenviable position of trying to catch up to the fantastic efforts of Grötschel-Holland and Padberg-Rinaldi, but we had the luxury of working alongside a broad and active worldwide community delving ever deeper into the theoretical side of the problem. Sifting through the growing body of research on the TSP would provide powerful tools for use in a computational attack. Before getting into the process, however, we made the single most important step in the overall effort, recruiting to our team David Applegate and Robert Bixby, two of the strongest computational mathematicians of our time. Things started slowly and we had several false starts, but in 1992 we solved a record 3,038-city drilling problem, utilizing a large network of computers working in parallel. With the pieces now in place, the team computed an

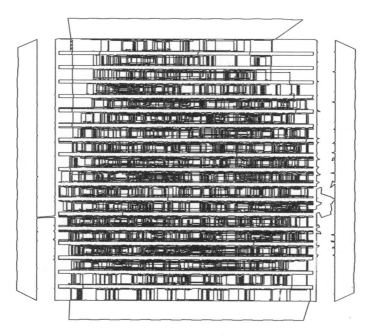

Figure 1.7
Solution of an 85,900-city TSP arising
in a computer-chip application.

Figure 1.8
Close-up view of a portion of the
85,900-city tour.

optimal 13,509-city tour through the United States in 1998, an optimal 24,978-city tour of Sweden in 2004, and, finally, an optimal tour for an 85,900-city applied instance in 2006. The computer code used in these solutions is called *Concorde* and it is available over the internet.

The 85,900 cities in the record problem represent locations of connections that must be cut by a laser to create a customized computer chip. The TSP in this case models the movement of the laser from location to location. Although movements are measured in fractions of an inch, the total travel time was a major contributor to the chip's production cost. The optimal route for the laser is illustrated in figure 1.7, with a close-up view of a small region in figure 1.8.

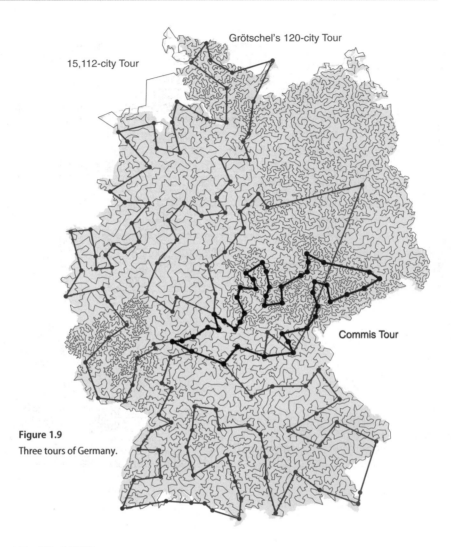

Grötschel's 120-city Tour

15,112-city Tour

Commis Tour

Figure 1.9
Three tours of Germany.

The World TSP

The grid-like distribution of points evident in the 85,900-city example, and in some of the drilling problems, does not really capture the traveling spirit of the 48-states tour that started the long TSP research program. But it is easy to appreciate the increased complexity of modern solutions by examining the three tours through Germany illustrated in figure 1.9. The small 33-city *Commis* tour was described in an 1832 book on tips for salesmen; the blue tour is Grötschel's 120-city record; and the tour in the background is an optimal route through 15,112 cities, computed with Concorde in 2001.

The 15,112-city route may be the final tour of Germany, but for an ultimate traveling challenge we put together a 1,904,711-city problem consisting of every city, town, and village in the world, including several research bases in Antarctica. Since 2001, this problem has withstood attacks by Concorde and by computer codes from around the globe. If the million-dollar Clay Prize is not to your taste, perhaps you would like to take on this World TSP Challenge. At the time of publication of this book, the best-known tour for the problem was produced by Danish computer scientist Keld Helsgaun. His tour of length 7,515,790,345 meters was found on October 10, 2010. This is almost certainly not the best-possible result, but we do know that no tour can be of length less than 7,512,218,268 meters, a bound computed with the Concorde code. Thus Helsgaun's tour is no more than 0.0476% longer than an optimal tour. That is close, but there is plenty of room for shortcuts.

Drawing the *Mona Lisa*

An optimal tour for the World TSP would be fantastic, but we are very likely more than a decade away from having the tools needed to make a serious attempt at its solution. Fortunately, there is no shortage of interesting problems to tackle along the way to conquering the world. A pretty example is the 100,000-city *Mona Lisa* TSP displayed in figure 1.10.

Figure 1.10
Leonardo da
Vinci's *Mona Lisa*
as a TSP. Tour found
by Yuichi Nagata.

This data set was developed in February 2009 by Robert Bosch, to create a continuous-line drawing of da Vinci's famous portrait. The current best Mona Lisa tour was found by Yuichi Nagata of the Japan Advanced Institute of Science and Technology. His tour is known to be at most 0.003% longer than an optimal solution. This is tantalizingly close, but we are not yet home. As an incentive to anyone who might want to weigh in on this problem, there is a $1,000 prize offered to the first person who can improve on Nagata's tour. A nice trophy, but keep in mind that the real goal of problem-by-problem challenges is to gather ideas for use in general solution methods for the salesman, and beyond to application areas well outside the TSP. New avenues of attack are the name of the game.

Road Map of the Book

The T-shirt displayed in figure 1.11, with artwork by Jessie Brainerd, a 2007 Budapest Semester in Mathematics student, would be interpreted immediately as the TSP by every recent graduate of applied mathematics or computer science who is worth his or her salt.[14] Study of the salesman is a rite of passage in many university programs, and short descriptions have even worked their way into recent texts for middle school students.

With the existing wide coverage of the problem, what am I trying to accomplish with this book? The answer is simple: I plan to take the reader on a path that goes well beyond basic familiarity of the TSP, moving right up to current theory and state-of-the-art solution machinery. The ultimate goal is to encourage readers to take up their own pursuit of the salesman, with the hope that a knockout blow will come from an as yet unknown corner.

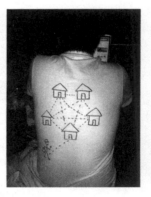

Figure 1.11
The TSP on Halloween 2007.
Photograph courtesy of Jessie
Brainerd.

We begin in chapter 2 by examining the roots of the salesman problem from both the mathematical and applied perspectives; the presentation of TSP history allows us to introduce basic themes picked up in later chapters. This is followed, in chapter 3, by a selection of the many applications of the TSP, including trip planning, genome sequencing, planet finding, and music arranging.

The heart of our technical treatment of the problem is the material presented in chapters 4 through 7, followed by a discussion of how TSP computer codes stack up to the task of solving large examples in chapter 8.

The $1,000,000 theoretical issue of a polynomial-time general method for the TSP is presented in chapter 9. If cold cash is what you desire, this is the chapter for you. I do not, however, recommend jumping ahead, even if your bank account is in desperate need of deposits. Indeed, the seeds of a successful theoretical attack may well be in methods that have proved themselves in the computational field of play. And if you are going for an impossibility result, you will need to handle the successful practical techniques in your proof.

Moving away from direct mathematics, in chapter 10 we cover studies on how humans, unaided by computers, go about solving the TSP; this area brings the problem into the realm of psychologists and neuroscientists. In chapter 11 we turn to the adoption of TSP tours in works of art, from the beautiful abstract paintings of Julian Lethbridge to the Jordan curves of Robert Bosch. Finally, chapter 12 wraps things up with a call for readers to take up the TSP challenge.

Figure 1.12
Left: W. Cook, far left, and V. Chvátal, far right, presenting author J. P. Donleavy a chamber pot, 1987. Photograph by Adrian Bondy. All rights reserved. Right: Postcard from J. P. Donleavy, 1987.

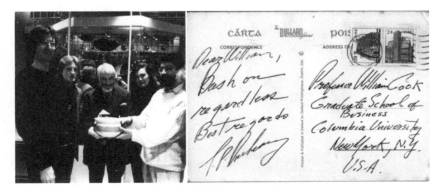

Bashing on Regardless

A bit of advice. When faced with an overwhelming number of slings and arrows, Irish writer J. P. Donleavy's character Rashers Ronald would vow to "Bash on regardless."[15] This became the rallying cry of the computational study of the TSP by Applegate et al. I recommend the reader, too, adopt this attitude when approaching the problem. We will cover work of numerous experts who have made huge advances, but the TSP remains essentially open. A new point of view could be just what is needed to dramatically alter our ability to tackle the salesman.

2: Origins of the Problem

It appears to have been discussed informally by mathematicians at mathematics meetings for many years.
—George Dantzig, Ray Fulkerson, and Selmer Johnson, 1954.[1]

The traveling salesman problem is known far and wide, but the path it has taken to such mathematical prominence is somewhat obscure. For example, we cannot say for certain when the problem's lively name first came into use. Nevertheless, most of the story can be told, albeit with the help of an educated guess here and there. Its telling serves the useful side purpose of getting our TSP feet wet before jumping in with details of current attempts to crack the notorious problem.

Before the Mathematicians

As a practical matter, the TSP was tackled by humans long before it became a fashionable topic of study in mathematics. Our cave-dwelling elders no doubt solved small versions while out hunting and gathering, but likely without the aid of much in the way of long-term planning. In recent centuries, however, members of certain professions clearly did take advantage of carefully planned routes. An examination of their tours is a good place to begin our discussion.

Salesmen

Foremost among the route planers is the namesake of the TSP. Consider the list of cities given in the sheet displayed in figure 2.1. This item is part of the correspondence of salesman H. M. Cleveland in the year 1925.[2] Mr. Cleveland worked for the Page Seed Company, gathering orders for

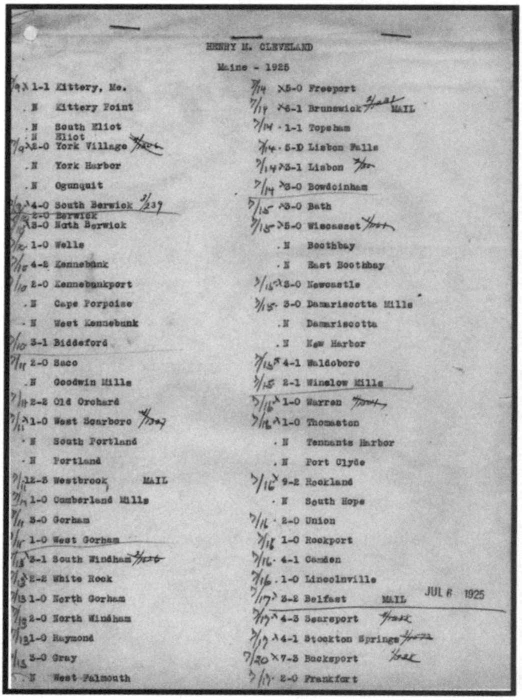
HENRY M. CLEVELAND

Maine - 1925

1-1 Kittery, Me.
. N Kittery Point
. N South Eliot
. N Eliot
2-0 York Village
. N York Harbor
. N Ogunquit
4-0 South Berwick
2-0 Berwick
3-0 North Berwick
1-0 Wells
4-2 Kennebunk
2-0 Kennebunkport
. N Cape Porpoise
. N West Kennebunk
3-1 Biddeford
2-0 Saco
. N Goodwin Mills
2-2 Old Orchard
1-0 West Scarboro
. N South Portland
. N Portland
12-3 Westbrook MAIL
1-0 Cumberland Mills
3-0 Gorham
1-0 West Gorham
3-1 South Windham
2-2 White Rock
1-0 North Gorham
2-0 North Windham
1-0 Raymond
3-0 Gray
. N West Falmouth
5-0 Freeport
6-1 Brunswick MAIL
1-1 Topsham
5-0 Lisbon Falls
3-1 Lisbon
3-0 Bowdoinham
3-0 Bath
5-0 Wiscasset
. N Boothbay
. N East Boothbay
3-0 Newcastle
3-0 Damariscotta Mills
. N Damariscotta
. N New Harbor
4-1 Waldoboro
2-1 Winslow Mills
1-0 Warren
1-0 Thomaston
. N Tennants Harbor
. N Port Clyde
9-2 Rockland
. N South Hope
2-0 Union
1-0 Rockport
4-1 Camden
1-0 Lincolnville
3-2 Belfast MAIL
JUL 6 1925
4-3 Searsport
4-1 Stockton Springs
7-3 Bucksport
2-0 Frankfort

Figure 2.1
Page Seed Company salesman list for Maine, 1925. One of five sheets.

corn and other products. His list of cities is one of five sheets outlining a tour of Maine. The full trip ran from July 9 through August 24, covering an amazing 350 stops.

Two observations make it clear that Mr. Cleveland and the Page Seed Company were interested in minimizing time spent on the road. First, the drawing of the tour, displayed in figure 2.2, reveals a remarkable efficiency in the itinerary; the portions where the tour appears to backtrack are all due to the available road network, where one town can only be reached by traveling to and from another town. Second, examine the following letter from Mr. Cleveland to his employer.

July 15, 1925

Dear Sirs

My route list is balled up the worst I ever saw. Will take half as long again to work it as last year. I have changed it some beginning with Stockton Springs, Frankfort, Winterport, Hampden Highlands, Bangor, Stillwater, Orono, Oldtown, Millford, Bradley,

Brewer, So Brewer, Orrington, So Orrington, Bucksport, then to original at Orland.

I wish you would send me my old list 1924 from Dexter on as it is much better than this. I don't see how you could break it out as you did especially from Albion to Madison would be jumping all over the map. This section I changed.

The river from Bangor down has no bridge and you have those towns down as if I could cross it anywhere. Last season's list was made out the best of any one and I can't see the object of changing it over. I think I have made myself plain.

—Yours truly, H. M. Cleveland

Mr. Cleveland was most unhappy with part of the tour and went ahead with his own improvements in its design.

Maine was just one of the destinations of Mr. Cleveland in 1925. He also traveled through Connecticut, Massachusetts, New York, and Vermont, making over 1,000 stops in total. And he was far from being the only

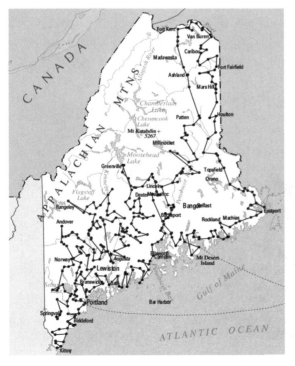

Figure 2.2
Page Seed Company salesman tour of Maine, 1925. The tour starts at Kittery and ends at nearby Springvale, both in the south of the state.

person making the rounds. Timothy Spears's book *100 Years on the Road: The Traveling Salesman in American Culture* cites an 1883 estimate by *Commercial Travelers Magazine* of 200,000 traveling salesmen working in the United States, and a further estimate of 350,000 by the turn of the century. This number continued to grow through the early 1900s, and in Mr. Cleveland's day the salesman was a familiar site in most American towns and villages.

Spears describes how these salesmen used aids such as L. P. Brockett's *Commercial Traveller's Guide Book* to map out routes through their regions. Often, however, tours were planned in a central office, such as was done in the Page Seed Company. The images in figure 2.3 indicate one way such tours were optimized, using pins and strings to plot potential routes on a map.

An important reference in this discussion is the 1832 German handbook *Der Handlungsreisende—Von einem alten Commis-Voyageur*.[3] The *Commis-Voyageur* describes the need for good tours.[4]

> Business leads the traveling salesman here and there, and there is not a good tour for all occurring cases; but through an expedient choice and division of the tour so much time can be won that we feel compelled to give guidelines about this. Everyone should use as much of the advice as he thinks useful for his application. We believe we can ensure as much that it will not be possible to plan the tours through Germany in consideration of the distances and the traveling

Figure 2.3
Rand McNally map cabinet and pin map. *Secretarial Studies*, 1922.

MAP CABINET   *Courtesy of Rand-McNally*

MAP SHOWING ROUTING OF SALESMEN BY PINS AND CORDS
. *Courtesy of Rand-McNally*

Figure 2.4
1832 German
salesman book.

back and forth, which deserves the traveler's special attention, with more economy. The main thing to remember is always to visit as many localities as possible without having to touch them twice.

This is an explicit description of the TSP, made by a traveling salesman himself!

The *Commis-Voyageur* book presents five routes through regions of Germany and Switzerland. Four of these routes include return visits to an earlier city that serves as a base for that part of the trip. The fifth route, however, is indeed a traveling salesman tour, indicated in figure 2.5. (The position of the route within Germany can be seen in the three-tours map displayed in figure 1.9.) As the *Commis-Voyageur* suggests, the tour is very good, perhaps even optimal, given road conditions at the time.

Numerous volumes written later in the century describe well-chosen routes in the United States, Britain, and other countries. The romantic image of the traveling salesman is captured, too, in stage, film, literature,

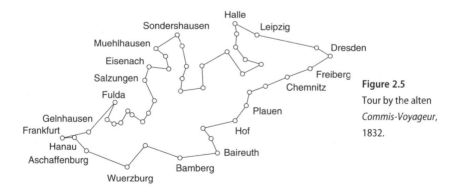

Figure 2.5
Tour by the alten
Commis-Voyageur,
1832.

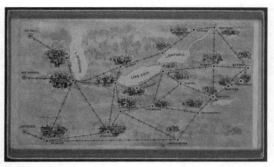

Figure 2.6
Commercial Traveller, McLoughlin Brothers,
1890. Courtesy of Pamela Walker Laird.

and song. The following is a typical turn-of-the-century salesman poem, taken from a compilation published in 1892.[5]

> Those who think a Drummer's life
> Is free from hardship, toil and strife,
> Are mistaken, for he has to go
> Through mud and rain, through sleet and snow.
> He sallies forth, gripsack in hand
> To seek for custom through the land.

The struggling drummer and his route-finding task were even featured in a board game, Commercial Traveller, created by McLoughlin Brothers in 1890, that asked players to build their own tours through a rail system. The choice of the salesman as the representative for the TSP is definitely well founded.

Lawyers

The salesman may have top billing, but other groups also traveled the land. The *Oxford English Dictionary* cites examples of the use of the word "circuit" as far back as the fifteenth century, concerning judicial districts in the United Kingdom. Traveling judges and lawyers served their districts by riding a circuit of the towns and villages, where court was held during specified times of the year. This practice was later adopted in the United States, where regional courts are still referred to as circuit courts, even though judges no longer take to the road.

Easily the best-known circuit rider in the history of the United States is the young Abraham Lincoln, who practiced law before becoming the country's sixteenth president. Lincoln worked the Eighth Judicial Circuit in the state of Illinois, covering fourteen county courthouses. His travel is described by Guy Fraker in the following passage.[6]

> Each spring and fall, court was held in consecutive weeks in each of the 14 counties, a week or less in each. The exception was Springfield, the state capital and the seat of Sangamon County. The fall term opened there for a period of two weeks. Then the lawyers traveled the fifty-five miles to Pekin, which replaced Tremont as the Tazewell County seat in 1850. After a week, they traveled the thirty-five miles to Metamora, where they spent three days. The next stop, thirty miles to the southeast, was Bloomington, the second-largest town in the circuit. Because of its size, it would generate more business, so they would probably stay there several days longer. From there they would travel to Mt. Pulaski, seat of Logan County, a distance of thirty-five miles; it had replaced Postville as county seat in 1848 and would soon lose out to the new city of Lincoln, to be named for one of the men in this entourage. The travelers would then continue to another county and then another and another until they had completed the entire circuit, taking a total of eleven weeks and traveling a distance of more than four hundred miles.

Fraker writes that Lincoln was one of the few court officials who regularly rode the entire circuit. A drawing of the route used by Lincoln and company in 1850 is given in figure 2.7. The tour is not quite the shortest possible (at least as the crow flies), but it is clear that it was constructed with an eye toward minimizing the travel of court personnel.

Preachers

The word circuit may have originated with the travel of judges and lawyers, but as a group they are rivaled in fame by the circuit-riding Christian preachers of the eighteenth and nineteenth centuries. John Hampson wrote the following passage in his 1791 biography of John Wesley, the founder of the Methodist church. "Every part of Britain and America is divided into regular portions, called circuits; and each circuit, containing twenty or thirty places, is supplied by a certain number of travelling preachers, from two to three or four, who go around it in a month or six weeks."[7] The conditions under which these men traveled is folklore in Britain, Canada,

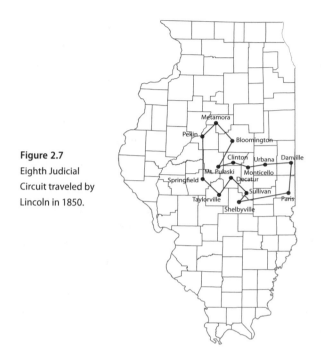

Figure 2.7

Eighth Judicial
Circuit traveled by
Lincoln in 1850.

and the United States. A feeling for the extent of their tours can be gathered
from the following quotes.

> I travelled about five thousand miles, preached about five hundred
> sermons, visited most of the circuits in Virginia and North Carolina.
> —Freeborn Garrettson, 1781.[8]

> Our circuit at that time, was five hundred miles around it, and for
> me to preach as I did sixty-three sermons in four weeks, and travel
> five hundred miles, was too hard. But I cried unto the Lord and he
> heard me; for as my day was, so was my strength.
> —Billy Hibbard, 1825.[9]

I have not been able to obtain detailed itineraries of any of the longer
circuits traveled by these Methodist preachers, but it is safe to assume that
some planning went into the selection of the routes. A goal of their work
was to reach as many church members as possible, so minimizing time on
the trail would have been an important consideration.

Euler and Hamilton

Back in the realm of mathematics, the plight of salesmen, lawyers, and preachers did not capture the attention of busy researchers, who had their hands full, laying down fundamentals for the rapidly expanding fields of the physical sciences. Two of the leading figures of the era did, however, explore aspects of the TSP, and they are rightly viewed as the grandfathers of traveling-salesman research.

Graph Theory and the Bridges of Königsberg

The great Leonhard Euler wrote the most important of all early mathematical papers describing touring problems. The Euler Archive cites an estimate by historian Clifford Truesdell that "in a listing of all of the mathematics, physics, mechanics, astronomy, and navigation work produced during the 18th Century, a full 25% would have been written by Leonhard Euler." History's most prolific mathematician studied a vast array of topics, including a puzzle that was a longstanding challenge to the residents of the town of Königsberg in East Prussia.

A satellite image of Königsberg, now called Kaliningrad, reveals the elaborate waterway formed by the River Pregel. The rectangular island created by the splitting of the river is called the Kneiphof; the large island to the east of the Kneiphof is called Lomse; the region north of the river is the Altstadt; and the region south of the river is the Vorstadt.[10]

In Euler's day, the Pregel was crossed by seven walkways: the Green and Köttel bridges joined the Kneiphof to the Altstadt, the Krämer and

Figure 2.8
Königsberg and the River
Pregel, TerraServer.com, 2011.

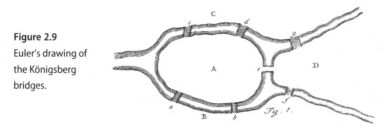

Figure 2.9
Euler's drawing of the Königsberg bridges.

Schmiede bridges joined the Kneiphof to the Vorstadt, the Honey bridge joined the Kneiphof and the Lomse, the High bridge joined the Lomse and the Altstadt, and the Wood bridge joined the Lomse and the Vorstadt. The good citizens of Königsberg enjoyed strolls through their town, crisscrossing the Pregel via the Green, Köttel, Krämer, Schmiede, Honey, Lomse, and Wood. The tale is that the Königsbergers had a standing challenge of crossing each of the seven bridges exactly once on a single walk through the town.

Euler weighed in on the Königsberg problem with a paper presented to the Academy of Sciences in Saint Petersburg on August 26, 1735.[11] His treatment follows a classic mathematical line of abstracting just the necessary information to capture the essence of the problem, and in so doing he laid the foundation for an important new branch of mathematics known as *graph theory*.[12]

To begin, Euler removed the physical nature of the challenge, sketching the town, river, and bridges, as displayed in figure 2.9. (This is a cleaned-up version of a computer scan taken from a copy of Euler's original published paper.) Euler labeled the regions of Königsberg as A, B, C, and D, and the seven bridges as a through g. These labels are enough to describe any route through the town, such as A to C via the c bridge, C to D via the g bridge, D to B via the f bridge, and B to A via the b bridge. A shorthand for this route would be $AcCgDfBbA$. Euler's arguments are based entirely on manipulating the routes as strings of symbols, rather than as walkers crossing the town.

The size of the land regions does not play a role in Euler's work, so the arguments can be visualized by a simple diagram, where A, B, C, and D are drawn as points, and a through g are drawn as lines joining pairs of these points, as in figure 2.10. The interpretation of the drawing is not influenced by the shape or length of the points and lines, but only by which pairs of points are joined. An object such as this is called a *graph*. The points of the graph are called *vertices*, the lines are called *edges*, and each edge has as its *ends* two of the vertices.

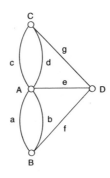

Figure 2.10
Graph representation of the Königsberg bridges.

In this stripped-down setting, a walk through Königsberg translates to movement from vertex to vertex in the graph, traveling along the graph's edges. A possible walk, starting at B and ending at D, is $B\,a\,A\,c\,C\,g\,D\,e\,A\,b\,B\,f\,D$. In this walk, three edges meet vertex B, namely a, b, and f; four edges meet vertex A, namely a, c, e, and b; two edges meet vertex C, namely c and g; and three edges meet vertex D, namely g, e, and f. The key observation of Euler is that the odd-even-even-odd pattern to these numbers is no accident: in any walk between two distinct points, the starting and ending vertices meet an odd number of edges and all other vertices meet an even number of edges. Furthermore, if we have a closed walk, that is, we start and end at the same point, then every vertex meets an even number of edges. So we have either all "even" vertices or exactly two "odd" vertices.

This is bad news for the Königsbergers. All four vertices of their bridge graph meet an odd number of edges, thus there can be no walk using each edge exactly once. Euler's short argument put an end to the Königsberg debate.

The Knight's Tour

Several years after settling the Königsberg puzzle, Euler wrote on a second touring challenge, known as the *knight's tour* problem in chess.[13] The task here is to find a sequence of knight's moves that take the piece from a starting square on a chessboard, through every other square exactly once, and then back to the starting square. Euler's solution is depicted in figure 2.11, where the order of moves is indicated by numbers on the squares.

The idea of a traveling knight appealed to Euler, who also laid out routes for boards of nonstandard size. These problems can be framed nicely using the language of graphs. In this case, we have a vertex for each square

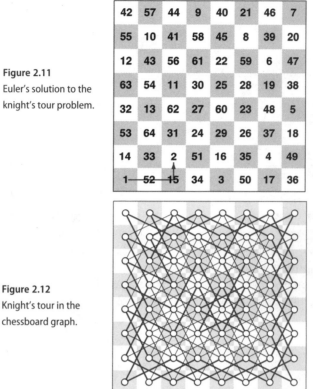

42	57	44	9	40	21	46	7
55	10	41	58	45	8	39	20
12	43	56	61	22	59	6	47
63	54	11	30	25	28	19	38
32	13	62	27	60	23	48	5
53	64	31	24	29	26	37	18
14	33	2	51	16	35	4	49
1	52	15	34	3	50	17	36

Figure 2.11
Euler's solution to the
knight's tour problem.

Figure 2.12
Knight's tour in the
chessboard graph.

on the board, with two vertices joined by an edge if a knight can travel
between the squares in a single move. A knight's tour is a closed walk that
visits each vertex exactly once. (Note the similarity with the Königsberg
problem, where we sought a walk traversing each edge exactly once.) The
particular graph for the full chessboard is displayed in figure 2.12, together
with Euler's route for the knight.

The Icosian

Ireland's Sir William Rowan Hamilton was also drawn to a question involving tours in a particular graph. A century after Euler, Hamilton studied
ways to visit all twenty corner points of the dodecahedron, the twelve-
sided Platonic solid. Hamilton made use of an abstract drawing he dubbed
the *Icosian*, displayed in figure 2.13. The lines of the Icosian represent the
dodecahedron's geometric edges and the circles represent its corners.

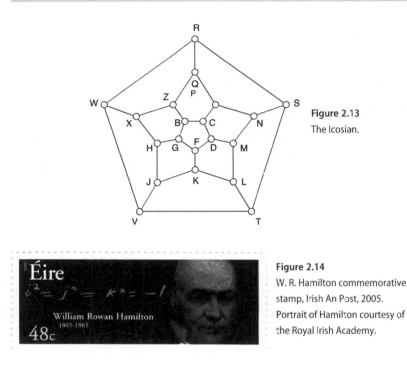

Figure 2.13
The Icosian.

Figure 2.14
W. R. Hamilton commemorative
stamp, Irish An Post, 2005.
Portrait of Hamilton courtesy of
the Royal Irish Academy.

The Icosian is a graph, and Hamilton's tours also proceed from vertex to vertex, always traveling along the graph's edges. Interestingly, Hamilton used an algebraic system to view this travel, in an approach similar in spirit to his defining equations for quaternions. He describes this in a formal letter to his friend John T. Graves in 1856.[14]

> As in the little paper which I lately sent you, let me continue to assume three symbols, i, κ, λ, which shall satisfy the four following equations:
>
> $$i^2 = 1, \ \kappa^3 = 1, \ \lambda^5 = 1, \ \lambda = i\kappa.$$
>
> What I have first to show, by one or two examples, is that the symbols so defined have curious but determinate properties, making them the legitimate instrument of a calculus: every symbolic result of which, so far as I can judge, and I have examined a great number of them, admits of easy and often interesting interpretation, with reference to the passage from face to face, or from corner to corner, of one or other of the solids considered in the ancient geometry.

The three symbols correspond to operations in the Icosian; when symbols are multiplied, the operations are made one after the other.[15] Through his calculus on these symbols, Hamilton showed that no matter what path

Figure 2.15
Left: The Icosian Game. Right: The Traveller's Dodecahedron.
© 2009 Hordern-Dalgety Collection, puzzlemuseum.com.

of five vertices is chosen as a start, it is always possible to complete a tour through the remaining vertices of the Icosian. Fascinated with this structure, Hamilton concluded his letter to Graves with a description of a game to be played on the Icosian graph.

> I have found that some young persons have been much amused by trying a new mathematical game which the Icosian furnishes, one person sticking five pins in any five consecutive points, such as *abcde*, or *abcde'*, and the other player then aiming to insert, which by the theory in this letter can always be done, fifteen other pins, in cyclical succession, so as to cover all the other points, and to end in immediate proximity to the pin wherewith his antagonist had begun.

Two versions of the game were later marketed by a British toy merchant. One variant, called the Icosian Game, consists of a wooden board with ivory pegs to mark visited points. The second variant, called the Traveller's Dodecahedron: A Voyage Round the World, is a handheld device, shaped as a partially flattened dodecahedron, with pegs for the points and a string to trace out the tour.[16]

Despite Hamilton's enthusiasm, the games were a flop in the commercial market. If you try a few rounds of play you will see quickly the problem: finding tours in the Icosian graph is too easy. Hamilton was quite defensive about this point, stating that the puzzles were not at all easy for him. This odd status, where the game is simple for children but challenging for Ireland's greatest mathematician, may have been due to Hamilton's algebraic view of things. Perhaps Hamilton was solving the puzzles through mental manipulation of i, κ, and λ, rather than tracing the tours visually.

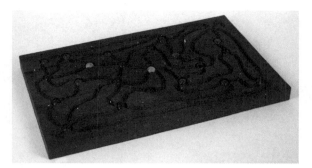

On a happier note, a twentieth-century variant of Hamilton's game did manage to bring in a significant number of sales. James Dalgety's Worried Woodworm puzzle, from 1975, asks for walks in a particular graph, but in this case the routes are tricky to spot. Dalgety's wooden board is displayed in figure 2.16. The main goal is to discover a path starting in the bottom left, ending at the top right, and visiting every hole along the way.

The Concorde code was used recently to settle additional Worried Woodworm challenges posed by Dalgety, but fair-minded players would, no doubt, frown on employing a state-of-the-art TSP solver and a high-powered computer to plot the worm's path through the twenty-three points.

Hamiltonian circuits

Euler's knights and Hamilton's game-playing children both search for tours in graphs, but what about a general question? Not all graphs possess a tour through their vertices and a challenge is to decide which do and which do not. Hamilton's fame added considerable luster to this challenge at a time when graph theory was just beginning to find its place within the mathematical world. This explains why his name gets top billing when describing the problem. But do not jump in alarm at the snub of Euler. Graph theorists reserve Euler's name for closed walks that model the sought-after trip through Königsberg. Thus, a *Hamiltonian circuit* in a graph is a closed walk that visits each vertex exactly once, while a *Eulerian walk* is a closed walk that travels along each edge exactly once. Both walks are fundamental concepts in graph theory, but there is a world of difference between the two, despite the obvious similarities.

Deciding whether or not a graph has a Hamiltonian circuit is an $\mathcal{NP}$-complete problem, capturing much of the complexity of the general TSP. On the other hand, there is a simple rule for determining if a graph has an

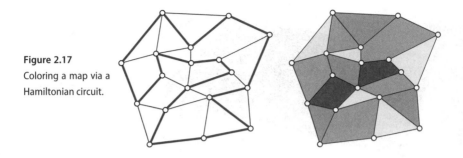

Figure 2.17
Coloring a map via a
Hamiltonian circuit.

Eulerian walk, namely, except for vertices meeting no edges at all, the graph must be connected, that is, it consists of a single piece, and each vertex must be the end of an even number of edges.

So we understand Euler, but not Hamilton. Indeed, year after year, brave mathematicians have suggested conditions guaranteeing Hamiltonian circuits, only to see their conjectures fail. A famous example is due to P. G. Tait in the 1880s. Tait was caught up in the excitement of Alfred Kempe's announced proof of the four-color theorem. This result states that the regions (countries) of any map can be colored with at most four colors in such a way that any two regions sharing a border receive different colors. Looking for an alternative proof that four colors suffice, Tait conjectured that a certain type of graph always has a Hamiltonian circuit.

To see the connection between traveling and map coloring, think of the boundaries of a map's regions as the edges of a graph, with the intersection points as vertices. A Hamiltonian circuit through this boundary graph gives a way to color the map, as illustrated in figure 2.17, where the red edges form a Hamiltonian circuit. Such a circuit does not cross itself, so it has an inside and an outside. Moreover, the border edges on the inside, that are not part of the circuit, cut across the inner area. We can thus color these inner regions with two colors, switching every time we cross one of the non-circuit edges. The same trick allows us to two-color the regions lying outside of the Hamiltonian circuit, yielding altogether a four-coloring of the map. In the example, the inner regions are colored dark yellow and light yellow, and the outer regions are colored dark blue and light blue.

Tait knew that not all maps have Hamiltonian circuits through their borders (the map of the continental United States is a ready example), but available tricks allowed the four-color problem to be restricted to maps such that each vertex of the border graph meets exactly three edges. Furthermore, the border graph could be assumed to be three-connected, that is, it is impossible to break the graph into two parts by deleting one or two vertices. Restricted to these three-regular, three-connected maps, Tait expected Hamiltonian circuits would always be available.

William Tutte, the great graph theorist and Bletchley Park code breaker, eventually showed Tait's conjecture to be false in 1946. That is too bad, but at least the circuit problem stood its ground longer than Kempe's four-color proof, which was shown to be incorrect by P. J. Heawood in 1890.

A historical footnote is that the first recorded description of the four-color problem is in a letter to Hamilton, written by Augustus De Morgan in 1852. Hamilton was not impressed with the problem, replying, "I am not likely to attempt your 'quaternion of colours' very soon."[17]

Mathematical Genealogy

Mathematicians enjoy tracing their academic heritage, following their Ph.D. thesis adviser to their adviser's adviser, and so on back through time. The Mathematics Genealogy Project Web site run by North Dakota State University contains over 130,000 records of Ph.D. advisers, with a goal to compile information on all the world's mathematicians. I am proud to trace my own roots back to Victorian-era mathematician Arthur Cayley, with an informal leap over to Sir Hamilton himself.

The path to Cayley is direct: W. Cook to U.S.R. Murty to C. R. Rao to Ronald Fisher to James Jeans to Edmund Whittaker to Andrew Forsyth to Arthur Cayley. The formal path stops here, since Cayley was trained in the law and did not obtain a Ph.D. degree. Cayley had great interest in mathematics, however, and in 1848 he traveled to Dublin to attend Hamilton's lectures on quaternions at Trinity College. Influenced by Hamilton, Cayley went on to write several hundred mathematical papers while practicing law, leading up to his appointment to the Sadleirian chair of mathematics at Cambridge in 1863. Cayley did not pick up Hamilton's interest in TSP-related problems, but he is a well-known figure in graph theory, introducing the notion of "trees" that we cover later in the book.

Vienna to Harvard to Princeton

Euler and Hamilton studied tours, but chessboards and dodecahedrons are a far cry from a salesman out on the road. A salesman is not satisfied with any old tour, she wants one of shortest possible length.

To bring in travel costs, we must jump ahead another century to Karl Menger and his work in Vienna. One of Menger's favorite topics in the 1920s was the study of techniques to measure lengths of curves in space.

This esoteric research likely provided inspiration for his announcement of a close relative of the TSP, made at a colloquium held on February 5, 1930.[18]

> We use the term *Messenger problem* (because this question is faced in practice by every postman, and, by the way, also by many travelers) for the task, given a finite number of points with known pairwise distances, to find the shortest path connecting the points.

The problem is to find a path through the points, without a return trip to the start. This is easily converted to the TSP by adding an extra "dummy" city that serves to link the ends of the path. The cost of travel between the dummy and each of the real cities can be set to zero, so that visiting the extra city will not influence the choice of the path's starting point or ending point.

Menger's "messenger problem" is recorded, in German, as part of the published documentation of the Vienna Mathematics Colloquium. The announcement is of clear historical importance, but it does not appear to have been the direct source of interest in the TSP among researchers in the United States. This honor goes to a lecture presented by prominent Harvard mathematician Hassler Whitney, cited in the following passage from Dantzig, Fulkerson, and Johnson's classic paper.[19]

> Merrill Flood (Columbia University) should be credited with stimulating interest in the traveling-salesman problem in many quarters. As early as 1937, he tried to obtain near optimal solutions in reference to routing of school buses. Both Flood and A. W. Tucker (Princeton University) recall that they heard the problem first in a seminar talk by Hassler Whitney at Princeton in 1934 (although Whitney, recently queried, does not seem to recall the problem).

Merrill Flood himself also credits Whitney's lecture when describing the history of the TSP in a 1956 research paper. "The problem was posed, in 1934, by Hassler Whitney in a seminar talk at Princeton University."[20] Even well after the fact, Flood refers to the TSP as the "48-states problem of Hassler Whitney" in an interview with Albert Tucker in 1984.[21]

It is natural to speculate on a possible connection between Menger's Vienna colloquium and Whitney's Princeton seminar. Support for such a connection was found by Alexander Schrijver, who notes that Menger and Whitney met at Harvard University in 1930–31, during a semester-long visit by Menger.[22] It is unclear, however, if the two actually exchanged information directly related to the salesman/messenger problem.

It also remains a question whether Whitney did in fact discuss the TSP at Princeton. There unfortunately is not an accessible record at

Figure 2.18
Hamiltonian circuit through Africa.

Princeton University covering the seminars delivered in the Department of Mathematics in the 1930s. The Pusey Library at Harvard University does, however, contain an archive of 3.9 cubic feet of Whitney's papers, and within the archive there is a set of handwritten notes that appear to be preparation for a seminar by Whitney, written sometime in the years shortly after 1930. The notes present an introduction to graph theory, including the following paragraph.

> A similar, but much more difficult problem is the following. Can we trace a simple closed curve in a graph through each vertex exactly once? This corresponds to the following problem. Given a set of countries, is it possible to travel through them in such a way that at the end of the trip we have visited each country exactly once?

In Whitney's problem, a graph is formed by placing a single vertex in each country, and joining two vertices by an edge if the countries share a border. A trip through the countries is a Hamiltonian circuit in the graph. This is an unusual choice as an example to describe the Hamiltonian-circuit problem and it is clearly not a far step from the TSP.

The geographic aspect of this example matches Flood's recollection of the "48-states problem." Indeed, Whitney's illustration of Hamiltonian circuits may well be the starting point of TSP research in the United States. In the words of Alan Hoffman and Philip Wolfe, Whitney served "possibly as a messenger from Menger" in bringing the salesman to the mathematics community.[23]

And on to the RAND Corporation

There is not a record of the study of the salesman problem, under the TSP name, in the late 1930s and into the 1940s, but by the end of the 1940s it had become a known challenge. At this point the center of TSP action had moved from Princeton to RAND, coinciding with Flood's relocation to California.

Princeton University's Harold Kuhn writes the following in a December 2008 e-mail letter.

> The traveling salesman problem was known by name around Fine Hall by 1949. For instance, it was one of a number of problems for which the RAND corporation offered a money prize. I believe that the list was posted on a bulletin board in Fine Hall in the academic year 1948–49.

The RAND prize list! The TSP literature is peppered with mention of these prizes, but it is no longer easy to track down a copy of the original RAND document. Hoffman and Wolfe describe the RAND prize as one "for a significant theorem bearing on the TSP." The list, together with the great reputation of the RAND research group, played an important role in spreading the news of the TSP, although the prize itself was never awarded.

Within RAND, a prize is mentioned by famed mathematician Julia Robinson, in remarks concerning her research into the theory of games. "And RAND was offering a $200 prize for its solution. In my paper, 'An iterative method of solving a game,' I showed that the procedure did indeed converge, but I didn't get the prize, because I was a RAND employee."[24] Likely inspired by another problem on the list, Robinson took up the study of the TSP in 1949. Her work on the salesman is in tune with a general approach to mathematics she describes in handwritten notes from this period. "I prefer working on problems whose statement is comparatively simple but where nothing is known about what sort of methods might lead to a solution, to working on those requiring extensions of existing methods."[25] The TSP certainly fit the bill—no progress on the problem was reported in the nearly twenty years since Menger's colloquium. As we will see in chapter 5, her contributions to the salesman provided the background for the RAND breakthrough several years later.

Whence the TSP?

In a 1949 research paper, Robinson uses "traveling salesman problem" in an offhand way, suggesting it was a familiar concept at the time. In fact, until a copy of the RAND prize list is uncovered from its likely hiding place in some archive or other, Robinson's report is the earliest known reference to the TSP by name. Robinson formulates the problem as finding "the shortest route for a salesman starting from Washington, visiting all the state capitals and then returning to Washington," matching both Flood's description and the data set used by Dantzig et al.

Robinson's language connects the TSP and the "48-states problem," but we do not know when and where the salesman name first came into play. Merrill Flood would seem to be the person to have this information, but unfortunately he does not, as he explained to Albert Tucker. "I don't know who coined the peppier name Traveling Salesman Problem for Whitney's problem, but that name certainly caught on, and the problem has turned out to be of very fundamental importance." Whatever the origin, except for small variations in spelling and punctuation, "traveling" versus "travelling," "salesman" versus "salesman's," etc., by the mid-1950s the TSP name was in wide use, and the problem was beginning to pick up its notorious reputation. The table was set for Dantzig et al.

A Statistical View

Many important problems in mathematics are attacked from all sides, sometimes without the attacking teams knowing others have joined the fray. Such is the case with the salesman problem. At about the time Flood and company were struggling with the TSP in the United States, on the other side of the world, statistician P. C. Mahalanobis took on the problem from a different mathematical point of view and with a far different application in mind.

Bengali Jute Farms

Mahalanobis is known as the Father of Statistics in India, founding both the Indian Statistical Institute and the *Sankhya* journal of statistics. One of his

Figure 2.19
Prasanta Chandra
Mahalanobis. Photograph
on right taken while on a
farm sample survey. Courtesy
Mahalanobis Museum,
Indian Statistical Institute,
Kolkata, India.

main interests was the development of techniques for carrying out sample surveys, and it is here he made a connection to the TSP.

A major source of revenue in India during the 1930s was obtained from its jute crop, accounting for roughly one quarter of total exports. The majority of India's jute was grown in the Bengal region and an important practical question was how to collect data to make accurate forecasts of the crop.

A complete survey of Bengali land used in jute production was impractical, owing to the fact that jute was grown on roughly six million small farms. Mahalanobis proposed instead to make a random sample survey, dividing the country into zones comprising land of similar characteristics, and within each zone selecting a random number of points to inspect for jute cultivation. A major component in the cost of making the survey would be the time spent in moving men and equipment from one sample area to the next. This is the TSP aspect of the application, to find efficient routes between the selected sites in the field. Concerning this, Mahalanobis writes the following in a 1940 research paper.[26]

It is also easy to see in a general way how the journey is likely to behave. Let's suppose that n sampling units are scattered at random within any area; and let's suppose that we may treat each such sample as a geometrical point. We may also assume that arrangements will usually be made to move from one sample point to another in such a way as to keep the total distance travelled as small as possible; that is, we may assume that the path traversed in going from one sample point to another will follow a straight line. In this case it is easy to see that the mathematical expectation of the total length of the path travelled in moving from one sample point to another will be $(\sqrt{n} - 1/\sqrt{n})$. The cost of the journey from sample to sample will therefore be roughly proportional to $(\sqrt{n} - 1/\sqrt{n})$. When n is large, that is, when we consider a sufficiently large area, we may expect the time required for moving from sample to sample will be roughly proportional to $\sqrt{n}$, where n is the total number of samples in the given area.

The term *expectation* refers to the average length of the optimal tours we would see if we repeated many times the experiment of taking n random points and solving the TSP. Perhaps owing to his research interests as a statistician, Mahalanobis does not discuss the operational task of actually finding tours for specific data. He focuses instead on making statistical estimates of the lengths of optimal routes. This is quite a different angle on the problem than that taken up by researchers at Princeton and RAND.

Mahalanobis's estimates were included in the projected costs of carrying out sample surveys in Bengal, and these projections were an important consideration in the decision to implement a small test in 1937 and a large survey in 1938.

Verifying the Tour Estimates

Mahalanobis did not give a precise analysis of his TSP formula, but his research set up a nice target for further work by the statistics community. The object of this work was to learn more about tours that arise when city locations are chosen at random in a unit square, that is, each point (x, y) with both x and y between 0 and 1 is equally likely to be chosen as a sample location. In particular, what can be said about the lengths of optimal tours through such point sets?

Researchers approached the problem from two directions. Eli Marks showed, in 1948, that the expected length of an optimal tour through a random set of points is at least

$$\frac{1}{\sqrt{2}} \left(\sqrt{n} - \frac{1}{\sqrt{n}} \right)$$

and M. N. Ghosh showed, in 1949, that the expected length is at most $1.27\sqrt{n}$. For large n, these results combine to prove that Mahalanobis's intuition was correct; the expected length of the tour is indeed proportional to $\sqrt{n}$.

In his research paper on the upper bound, Ghosh made a point to comment on the operational task of producing results for specific data. "After locating the n random points in a map of the region, it is very difficult to find out *actually* the shortest path connecting the points, unless the number n is very small, which is seldom the case for a large-scale survey."[27]

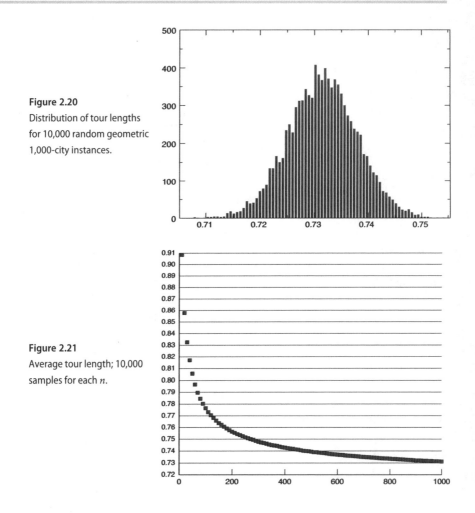

Figure 2.20
Distribution of tour lengths for 10,000 random geometric 1,000-city instances.

Figure 2.21
Average tour length; 10,000 samples for each n.

It is interesting that he observed the heart of the TSP challenge of finding optimal tours, apparently without connection to Menger, Whitney, and Flood.

The TSP Constant

The Mahalanobis-Marks-Ghosh result gives an estimate for the average tour length, but it does not say anything about the range of lengths we are likely to see in a series of experiments: some random point sets might have long optimal tours, while others could have tours that are quite short. This in fact does not happen, if n is reasonably large. To understand this point, examine the histogram given in figure 2.20, displaying the optimal tour lengths divided by $\sqrt{1,000}$ for 10,000 random geometric instances, each with 1,000 cities. The results form a nice bell curve around the mean 0.7313. With only 1,000 cities there is still some variance in the tour values, but a famous theorem of Beardwood, Halton, and Hammersley, published in 1959, implies that as n gets large, the distribution of tour lengths will spike around a particular number called β, the TSP constant.[28]

An intriguing question is to determine the value of β. Its investigation has led to an important subfield of probability, but proven estimates do not come close to pinning it down. So we have a natural constant whose actual value is unknown.

In an ongoing study of β with David Applegate, David Johnson, and Neil Sloane, we have solved over 600,000,000 geometric instances of the TSP. This has given Concorde quite a workout, but the mountain of computation alone cannot prove any definitive results. Nonetheless, plots such as the one displayed in figure 2.21 strongly suggest a steady decrease in the average tour length divided by $\sqrt{n}$ as n increases, pointing toward an ultimate value of approximately 0.712 for β.[29]

3: The Salesman in Action

Because my mathematics has its origin in a real problem doesn't make it less interesting to me—just the other way around.

—George Dantzig, 1986.[1]

The name itself announces the applied nature of the traveling salesman problem. This has surely contributed to a focus on computational issues, keeping the research topic well away from perils famously described in John von Neumann's essay "The Mathematician". "In other words, at a great distance from its empirical source, or after much 'abstract' inbreeding, a mathematical subject is in danger of degeneration". Indeed, a strength of TSP research is the steady stream of practical applications that breathe new life into the area.

Road Trips

In our roundup of TSP applications, let's begin with a sample of tours taken by humans, including the namesake of the problem.

Salesmen in the Digital Age

An automobile equipped with a global positioning system (GPS) device is the mode of transportation typically chosen by local traveling salesmen. Mapping software running on the GPS unit often includes a TSP solver for small instances having a dozen or so cities, and this is usually adequate for daily trips. Detailed maps stored in the unit can be used to deliver accurate estimates of the time to travel from point to point, allowing TSP solutions to reflect actual driving conditions faced by travelers.

Alain Kornhauser of Princeton University, an expert on the application of mapping technology, described an interesting, reverse, use of GPS equipment. When a user specifies a destination with a latitude and longitude, it is sometimes impossible to project the point onto the known grid of roads and highways—there just isn't a way to get to the location. But if a package must be delivered, then a local trucker will often find a route, perhaps using a small lane that is not on the grid. In such a case, the GPS system reports back to a central server and a link is added into the grid, tracing the path traveled by the vehicle. Next time a delivery is requested for the location, the mapping software makes use of the newly inserted road.

Pick-ups and Deliveries

A common use of small-scale TSP models is the routing of buses and vans to pick up and deliver people and packages. Merrill Flood wrote that a school bus–routing problem provided his initiation into the study of the TSP. Another early team, George Morton and Ailsa Land of the London School of Economics, was drawn to the problem by a laundry-van application. In a more recent example, the firm Rapidis employed Concorde to plot routes for their customer Forbruger-Kontakt, a distributor of advertising material and samples, operating in Denmark and several other countries. The image in figure 3.1 is a drawing made from a screen dump of the routing software created by Rapidis. The route in the drawing obeys one-way streets and other travel restrictions, making the cost to travel between two points depend on the direction that is chosen.

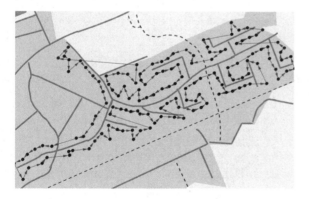

Figure 3.1
TSP tour for deliveries by Forbruger-Kontakt. Courtesy of Thomas Isrealsen.

Meals on Wheels

A team from Georgia Tech described a successful application of a fast TSP heuristic algorithm for constructing the routes of aid workers in a "Meals on Wheels" program in Atlanta.[2] Each driver in the program delivers meals to 30 to 40 locations out of a total of 200 or so that are served daily. To construct routes for the drivers, all 200 locations are placed in a tour that is divided into segments of the appropriate lengths. The overall tour is found with the aid of the spacefilling curve illustrated in figure 3.2. Ever-finer versions of the curve will eventually include any point in the city, and the heuristic tour through the 200 locations is obtained by taking the order in which the locations appear on the curve.

The simplicity of the tour-finding method allowed the manager of the program to easily update the tour by hand as new clients joined the system and existing clients left the system. The process runs as follows. The position of a point in the tour depends only on its relative position θ on the spacefilling curve. The Georgia Tech team precomputed the value of θ for a fine grid of (x, y) locations from a standard map of Atlanta. The list of active clients was stored on two sets of index cards, one sorted alphabetically and the other stored in the tour order, that is, by increasing value of θ. To delete a client, his two cards are simply removed. To insert

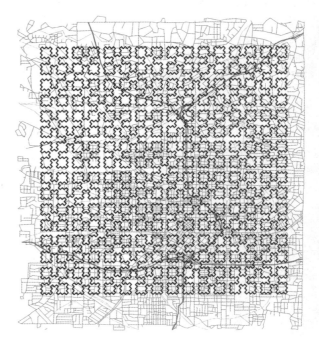

Figure 3.2
Spacefilling curve for Atlanta region. Image courtesy of John Bartholdi.

a new client, the map is used to determine the (x, y) coordinates of the client's location, the table is used to look up the corresponding value of θ, and θ is used to insert the client's card into the tour order. An ingenious, low-tech solution for a practical TSP application.

Farms, Oil Platforms, and Blue-claw Crabs

The farming study of Mahalanobis in the 1930s is an early example of the use of the TSP in planning inspections of remote sites. This type of logistical application occurs in many other contexts as well. For example, William Pulleyblank reports the use of TSP software to plan routes for an oil firm to visit a set of 47 platforms off the coast of Nigeria. In this instance, the platforms are visited via a helicopter flying from an onshore base. In another example, a group at the University of Maryland modeled the problem of scheduling boat-crew visits to approximately 200 stations in the Chesapeake Bay. The purpose of the boat trips was to monitor the blue-claw crab population in the bay; the researchers turned to the TSP after having difficulty completing trips quickly enough to permit frequent monitoring of all sites.

Book Tours

Manil Suri, the author of the novel *The Death of Vishnu* and a professor of mathematics, made the following remark in *SIAM News*.[3]

> The initial U.S. book tour, which starts January 24, 2001, will cover 13 cities in three weeks. When my publisher gave me the list of cities, I realized something amazing. I was actually going to live the Traveling Salesman Problem! I tried conveying my excitement to the publicity department, tried explaining to them the mathematical significance of all this, and how we could perhaps come up with an optimal solution, etc., etc. They were quite uneasy about my enthusiasm and assured me that they had lots of experience in planning itineraries, and would get back to me if they required mathematical assistance. So far, they haven't.

Despite the reluctance of Suri's publishers, book touring is a natural setting for the TSP.

Extra Miler Club

The motto of the Extra Miler Club is "because the shortest distance between two points is no fun." Nonetheless, members do like to plan their tours, aiming to visit all 3,100+ counties in the United States. This is not exactly a TSP, since crossing any point of a county line is sufficient, although some members prefer to visit each county seat of government.

The *Wall Street Journal* reported that one Extra Miler proposed to eat a Big Mac in each of the over 13,000 McDonald's in North America.[4] That would be a nice application of the TSP, but the club Web site reports the member "has now set forth upon a less gastronomically challenging goal."

The Iron Butt Rally

While the Extra Milers typically travel by automobile, the vehicle of choice for the 35,000-member strong Iron Butt Association is the motorcycle. One of their many challenges is the 48 States in 10 Days ride, where riders must visit all 48 continental states in the United States. Any route through the states is acceptable, but riders must obtain printed documentation, such as a gasoline receipt, verifying each state on their trip. Rider Maura Gatensby sent an e-mail letter in February 2009, asking about the city locations used in the Dantzig-Fulkerson-Johnson TSP tour.

> Most of us work on the problem by taking existing routes and trying to trim them somehow, but after reading about the existence of this problem in mathematics, I would like to make the Dantzig route my base route, and then perhaps try and reduce distance by moving some of the Dantzig locations. If the Dantzig route is not too long, I would like to just ride this route, because of its historical significance. Sometimes the shortest route isn't the "best" route, there is more poetry in walking in the footsteps of giants.

This is certainly a great use of their optimal solution.

Ms. Gatensby writes that the shortest distance known for the 48/10 ride is 6,967 miles. Although 48 cities is easy for today's TSP solvers, the problem is complicated by the fact that there are many choices for potential stops in each state. It would be an interesting challenge to find the optimal route with some fixed constraint, such as requiring each state visit to be among a list of known gasoline stations.

Figure 3.3
En route with the *Miss Izzy*.
Photograph courtesy
of Ron Schreck.

Flight Times

For record speed, it is hard to beat Ron Schreck, who uses the RV-8 airplane *Miss Izzy* for his tours. In 2007, Schreck had the idea to visit in a single day all 109 public airports in his home state of North Carolina. Concorde provided an optimal tour that Ron modified slightly to reach before sunrise several airports having lighted runways. His trip was made on July 4, a public holiday in the United States, which helped in avoiding delays. Schreck's total flight covered 1,991 miles in seventeen hours, with the time between landings averaging only nine and a half minutes. Landing here typically meant touching the wheels on the ground and bouncing back into the air.

Mapping Genomes

Turning away from the movement of people and vehicles we find surprising uses of the TSP model. One of the most interesting of these arises in genetics research, where a focus over the past decade has been the accurate placement of *markers* that serve as landmarks for genome maps.

A genome map has for each chromosome a sequence of markers with estimates of the distances between adjacent markers. The markers in these maps are segments of DNA that appear exactly once in the genome and can be reliably detected in laboratory work. The ability to recognize these unique segments allows researchers to use them to verify, compare, and combine physical maps created across different laboratories. It is particularly useful to have accurate information on the order in which the markers appear on the genome, and this is where the TSP comes into play.

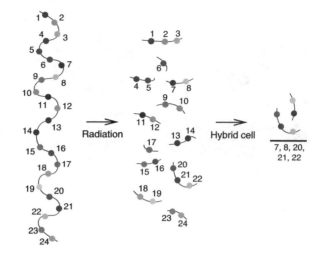

Figure 3.4

Radiation hybrid mapping.

One of the primary techniques for obtaining laboratory data on the relative position of markers is known as *radiation hybrid* (RH) *mapping*. This process exposes a genome to high levels of X-rays to break it into fragments. The fragments are then combined with genetic material, taken from rodents, to form hybrid cell lines that can be analyzed for the presence of markers. A simple illustration of the two steps is given in figure 3.4.

The central theme in RH mapping is that positional information can be gleaned from an analysis of which pairs of markers appear together in cell lines. If two markers A and B are close on the genome, then they are unlikely to be split apart in the radiation step. Thus, in this case, if A is present in a cell line, it is likely that B is present as well. On the other hand, if A and B are far apart on the genome, then we can expect to have cell lines that contain just A or just B, and only rarely a cell line containing both A and B. This positional reasoning can be crafted into a notion of an experimental distance between two markers.

Using the experimental distances, the problem of finding the genome order can be modeled as a TSP. Indeed, a genome ordering can be viewed as a path traveling through each marker in the collection. As usual, such a Hamiltonian path is readily converted to a tour by adding an extra city to permit the ends of the path to be joined.

A group at the National Institutes of Health (NIH), led by Richa Agarwala and Alejandro Schäffer, has developed methods and software for handling these genome TSP problems in practical settings, including procedures for dealing with erroneous data (a common occurrence in laboratories).[5] The NIH package uses Concorde to permit the software to find optimal tours; it has been adopted in a number of important studies, including the construction of human, macaque, horse, dog, cat, mouse, rat, cow, sheep, and river buffalo maps.

Aiming Telescopes, X-rays, and Lasers

Although we normally associate the TSP with applications that require physical visits to remote locations, the problem also arises when sites can be observed from afar, without actual travel. A natural example is when the sites are planets, stars, and galaxies, and the observations are to be made with some form of telescope.

The process of rotating equipment into position to make an observation is called *slewing*. For large-scale telescopes, slewing is a complicated and time-consuming procedure, handled by computer-driven motors. In this setting, a TSP tour that minimizes the total slewing time for a set of observations can be implemented as part of an overall scheduling process. The cities in the TSP are the objects to be imaged and the travel costs are estimates of the slewing times to move from one object to the next.

In a *Scientific American* article, Shawn Carlson describes how a TSP heuristic came to his aid in scheduling a fragile, older telescope to image approximately 200 galaxies per night. Concerning the need for good TSP tours, Carlson writes the following. "Because large excursions from horizon to horizon sent the telescope's 40-year-old drive system into shock, it was vital that the feeble old veteran be moved as little as possible".[6] Modern telescope installations are certainly not feeble, but good solutions to the TSP are vital for the efficient use of very costly equipment.

Finding Planets

Interesting examples of the TSP have been considered in planning work for space-based telescope missions by NASA. Martin Lu of the Jet Propulsion Laboratory calls this study the "traveling planet-finder problem" since a major goal is the discovery of Earth-like planets in orbit around nearby stars.

As in the case of ground-based equipment, the TSP is used to determine the sequence of observations to be made by the telescopes. In this setting, however, the sequencing of observations is made well in advance of the mission, rather than on a nightly basis. This preplanning is due to the great amount of fuel consumed in slewing operations and to the length of time needed to study each star. Martin Lu estimates that approximately fifty stars would be observed in a three-year mission.

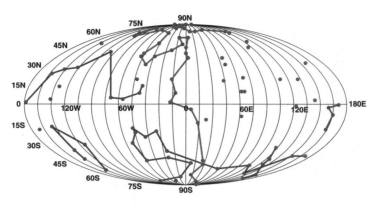

Figure 3.5
Two-occulter formation visiting 80 stars, by E. Kolemen and N. J. Kasdin.

A difficulty with observing a possible Earth-like planet is that light directly from its star washes out any image of the planet itself. A proposed solution consists of using a space telescope together with a large occulter stationed 50,000 to 100,000 kilometers away. Robert Vanderbei of Princeton University describes this as holding a giant thumb in front of the telescope's eye to block out starlight. The telescope remains in a fixed orbit, while the occulter moves from one position to another, setting up the observations.

A detailed study of the sequencing of occulter-based telescopes has been carried out at Princeton University by Egemen Kolemen and Jeremy Kasdin.[7] They use a sequence of optimization models to estimate fuel costs in moving the occulter from star to star. The image in figure 3.5 depicts their solution when a single telescope works with two occulters, alternating the observations from one to another; the colored paths represent the tours taken by the occulters. Note that sub-paths that appear to be isolated in the figure are actually connected around the back of the sphere. In this test, the solution picks out 80 of the top 100 candidate stars in a NASA target list.

X-ray Crystallography

The ground-based telescope TSP is similar to a study by Robert Bland and David Shallcross in a different domain.[8] Working with a team at Cornell University in the mid-1980s, Bland and Shallcross used the TSP

Figure 3.6
Crystal image drawn by laser.

to guide a diffractometer in X-ray crystallography. The travel costs in this case are estimates of the time for computer-driven motors to reposition sample crystals and to aim the X-ray equipment; experiments can consist of up to 30,000 observations per crystal. Bland and Shallcross reported improvements of up to 46% in total slewing time with the help of TSP methods.

Lasers for Crystal Art

The use of pulsed lasers in manufacturing settings provides another opportunity for this type of "aiming" TSP. A nice example is in the production of models and artwork burned into clear solid crystals, such as the pla85900 object produced by Mark Dickens of Precision Laser Art, displayed in figure 3.6. The focal point of a laser beam is used to create fractures at specified three-dimensional locations in the crystal, creating tiny points that are visible in the clear material. The TSP is to guide the laser through the points to minimize production time.

Dickens has adopted heuristic methods from Concorde to handle very large sets of points needed to obtain high-quality reproductions of elaborate images. This application holds a place of honor as having generated the largest industrial instances of the TSP we have encountered to date, with some examples exceeding one million cities.

Guiding Industrial Machines

In modern manufacturing, machines are often adopted to perform repeated tasks, such as drilling holes or attaching items. This is a common setting for TSP applications.

Figure 3.7
Printed circuit board with 441 holes. Photograph courtesy of Martin Grötschel.

Drilling Circuit Boards

Printed circuit boards contained in common electronic devices often have numerous holes for mounting computer chips or for making connections between layers. The holes are produced by automated drilling machines that move between specified locations to create one hole after another, and a classic application of the TSP is to minimize the travel time of the drill head during the production process. Gerhard Reinelt's TSPLIB test set contains a number of examples of this type, including an instance based on the board displayed in figure 3.7.

The use of TSP algorithms has led to improvements of approximately 10% in the overall throughput of circuit-board production lines.[9] Typical problems in this class range in size from several hundred cities up to several thousand cities.

Soldering a Printed Circuit Board

Wladimir Nickel, an electronics engineer in Germany, wrote that he has adopted Concorde in a follow-up step in circuit-board production, where items are soldered onto the surface of the board. He uses a computer numerical controlled (CNC) machine, equipped with a solder paste dispenser, to print solder at specified locations. His machine is displayed in the photographs given in figure 3.8; the board being created has 256 solder locations and the TSP solution provides the quickest way to move the dispenser through the full set of points.

Figure 3.8
Applying solder to a printed circuit board. Courtesy of Wladimir Nickel.

Engraving Brass

Brass dies are used in printing raised images, such as those found on boxes of chocolates. The dies were once made by hand, but now they are typically engraved with heavy-duty CNC milling machines. When a CNC machine has completed the cutting of a letter or design element, the spindle is raised and the device moves to the next letter or element. Additional flexibility in this case comes from the fact that elements to be cut are not single points, so the machine can be guided to any location above the element. CNC engraver Bartosz Wucke wrote in 2008 that the application of the TSP reduced the working time by half in cases where dies have significant amounts of text or where there are abstract patterns of many points.

Customized Computer Chips

The same class of application, on a much smaller physical scale, arose in work at Bell Laboratories in the mid-1980s. Bell researchers developed a technique for the quick production of customized computer chips. The process starts with a basic chip having a network of simple building blocks, called logic gates. Portions of the network are then cut with a laser to create individual groups of gates that allow the chip to perform some described function. In this case, the TSP is to guide the laser through the locations that need to be cut. Jon Bentley and David Johnson provided fast TSP heuristic methods that lowered the slewing time by over 50% on typical examples, providing an important speedup in the production process.

   This application also holds a place of honor as the source of the record 85,900-city TSP instance displayed in figure 1.7.

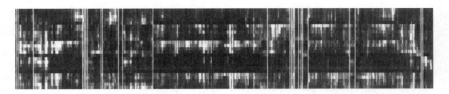

Figure 3.9
Gene expression data. Image courtesy of Sharlee Climer and Weixiong Zhang.

Cleaning Silicon Wafers

Another TSP application arises earlier in the production of computer chips. Standard chips are etched into large circular wafers of silicon and these wafers must be free of all impurities. The nanomanufacturing firm Applied Materials has a technique for cleaning defects on wafers and they have used Concorde to guide the machinery from one defect to another.

Organizing Data

Organizing information into groups of elements with similar properties is a basic tool in data mining, the process of extracting patterns from data. The TSP has been adopted in such efforts when there is a good measure of the similarity between pairs of data points. Using the similarity values as travel costs, a Hamiltonian path of maximum cost places similar points near to one another (since closely related points have high similarity measures), and thus segments in the path can be used as candidates for clusters. The final splitting into segments is typically done by hand, selecting natural breakpoints in the ordering.[10]

An elegant alternative to this two-stage method was proposed by researchers Sharlee Climer and Weixiong Zhang.[11] In their approach, $k + 1$ dummy cities are added when creating the TSP, rather than just a single city. Each of the dummy cities is assigned a travel cost of zero to all other cities. The additional cities serve as breakpoints to identify k clusters, since a good tour will use the zero-cost connections to dummy cities to replace large travel costs between clusters of points.

Climer and Zhang use their TSP+k method as a tool for clustering gene expression data, adopting Concorde to compute optimal tours and varying k to study the impact of different cluster counts. The image in figure 3.9 was produced with their software. The data set displayed in the figure consists of 499 genes from the plant *Arabidopis* under five different environmental

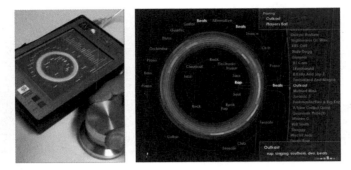

Figure 3.10
The MusicRainbow
device. Courtesy of
Elias Pampalk.

conditions; the shades of gray represent gene expression values; the clusters are indicated by solid white lines.

Musical Tours

The TSP has also been used to make sense out of vast collections of computer-encoded music. Elias Pampalk and Masataka Goto, working out of the National Institute of Advanced Industrial Science and Technology in Japan, created the *MusicRainbow* system to support users in discovering new artists that may appeal to their musical tastes. Pampalk and Goto took a collection of 15,336 tracks from 558 artists and developed a similarity measure between each pair of artists, computed by comparing audio properties of the tracks in the collection. The TSP was then used to arrange the artists in a circular order, such that similar artists are near to one another. In this application the cities are the musicians and the travel costs are the similarity measures.

Using the circular ordering, the music collection can be navigated by turning a knob, with artist information displayed on a computer screen. Various identifiers associated with the artists are indicated via a set of concentric colored rings corresponding to high-level classifications, such as rock and jazz. A nice feature of MusicRainbow is that all identifier information is obtained automatically via a search for Web pages, allowing the system to be easily deployed on any music collection.

Elias Pampalk was involved in a second music-related TSP application, together with colleagues Tim Pohle and Gerhard Widmer from the University of Linz in Austria. The idea this time is to organize a collection of music tracks into a circular list, such that similar pieces are near to one another. Such an arrangement allows a user to spin a wheel to pick a piece suiting their current mood, and the player follows this with a sequence of similar tracks. In their *Traveller's Sound Player* demonstration, the team used timbral similarities to measure the distance between pairs of tracks.

A test case included over 3,000 tracks, and the TSP was used to minimize the total distance in the circular order.

On a more local level, New York University's Drew Krause adopts the TSP as an aid in creating individual compositions. Smooth transitions in music are called conjunct melodies, and they are associated with pleasing sound. In Krause's process, Concorde is used to build arrangements with minimum transitions from one chord to the next; the cities are a collection of chords and the travel costs are defined as the sum of the half-step distances between the corresponding notes.

Speeding Up Video Games

Modern video games use large amounts of data to give objects in their displays an appearance of physical material, such as wood or metal. The basic components of this display data are called *textures* and libraries of thousands of textures are available, ranging from bricks to rust. Any scene in a game requires a specific set of textures to render the displayed objects, and a challenge is to get the texture data onto the video monitor as quickly as possible, to give smooth transitions from scene to scene. This is where the TSP can help.

A basic property of data access on digital video disks (DVD) is that reading items stored sequentially is much faster than accessing items from random locations. It follows that the layout of texture data on a disk can have a large impact on the time needed to render a game scene. It is highly desirable to have sets of textures used in the game residing sequentially, but this is typically not possible unless textures are duplicated on the disk, greatly increasing the storage requirement. As an alternative, the layout can be chosen such that the total number of breaks is minimized, where a break occurs when a set of textures needed in the game is stored in more than one location on the disk. If a texture set is split into k intervals, then it contributes $k - 1$ breaks to the layout. This is the same measure used in the genome-mapping application, where texture sets correspond to cell lines. In this TSP setting, the cities are the textures and the cost of travel between two textures is the number of sets that contain one of the textures but not the other. Like the genome problem, this application calls for a Hamiltonian path rather than a tour, which is handled in the usual manner via the addition of an extra city.

This application was described by Glen Miner of the Canadian firm Digital Extremes. Digital Extremes has experimented with the Concorde code for producing texture layouts, reporting significant improvements through the use of the TSP.

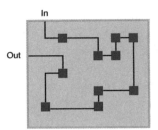

Figure 3.11
Scan chain.

Tests for Microprocessors

The computing technology firm NVIDIA recently adopted Concorde in optimizing the on-chip circuitry used to test their graphics processors. This is a common use of the TSP in the design of modern computer chips, where post-manufacturing testing is a critical step in the production process.

To facilitate such testing, *scan chains* were introduced in the 1980s to link components, or scan points, of a computer chip in a path having input and output connections on the chip's boundary, as illustrated in figure 3.11. A scan chain permits test data to be loaded into the scan points through the input end, and after the chip performs a series of test operations the data can be read and evaluated at the output end.

The TSP is used to determine the ordering of scan points to make the chain as short as possible. Minimizing the chain length helps to meet a number of goals, including saving valuable wiring space on the chip and saving time in the testing phase by allowing signals to be sent more quickly.

In most cases chip manufacturing technology allows only horizontal and vertical connections, thus the distance between two points in a scan-chain TSP is measured using paths that travel only horizontally or vertically, such as walking the streets of Manhattan. A drawing of an optimal path for a 764-city scan-chain problem is given in figure 3.12. This example

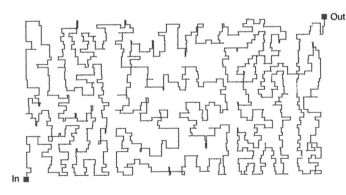

Figure 3.12
Scan-chain TSP
with 764 cities.

was provided by Michael Jacobs and Andre Rohe of Sun Microsystems and it was solved using the Concorde code. To reduce the time required for testing, a modern computer chip will typically have multiple scan chains; the 764-city example was one of twenty-five chains on the given chip.

Scheduling Jobs

The German firm BÖWE CARDTEC delivers hardware and software products for managing the production of smart cards, such as credit cards and identification cards. Their customers typically produce many types of cards on the same hardware and this requires reconfiguration steps between production runs, such as a change in the ribbon color and the insertion of the correct blank cards. The setup time between different jobs is significant and reduces the overall daily production. To address this, BÖWE CARDTEC software uses the TSP to sequence jobs in an order that minimizes the total setup time: the cities are the jobs and the travel cost between jobs i and j is the time it takes to reconfigure the machine for job j after it has completed job i. The firm reports that using tours obtained with Concorde reduced the total setup time by up to 65% in typical applications, resulting in significant gains in the overall rate of production.

This type of scheduling application was first described by Merrill Flood in a lecture given in 1954. In typical examples, the setup time to move from job i to job j is different than the time to move from job j back to job i. The TSP thus takes the asymmetrical form, where the cost of a tour depends on the direction of travel.

And More

The areas of application we have described by no means exhaust the reach of the traveling salesman. Indeed, intriguing new uses for the model appear regularly in the applied mathematics literature. Successful projects that have been reported include the following:[12]

- planning hiking paths in a nature park
- minimizing wallpaper waste
- picking items in a rectangular warehouse
- cutting patterns in the glass industry
- constructing universal DNA linkers
- estimating the trenching costs for connecting a telescope array

- studying problems in evolutionary change
- assembling a genome map from a library of known subsequences
- gathering geophysical seismic data
- compressing large data sets of zero-one-valued arrays

Some of these settings are wildly distant from actual salesmen planning their tours.

4: Searching for a Tour

We do not claim that our program is infallible, but rather that it gives good answers in a computationally feasible amount of computer time.

—Robert Karg and Gerald Thompson, 1964.[1]

A salesman on the road will not be impressed by a claim of TSP unsolvability. She will nonetheless start up the car and get on with the task of visiting customers. This practical mind-set argues for an alternative approach to the problem: let's give up for now the notion that only the absolute best solution will do, and focus on delivering, as quickly as possible, a near-optimal route. Such a view opens the door to all sorts of creative ideas for getting the salesman home in time for dinner. Indeed, some of the techniques developed and employed in this branch of TSP research are now workhorses in computational science, such as simulated annealing, genetic algorithms, and local search. Tour finding serves as a sandbox for testing methods that aim to select a good solution from a large population. It is the playground of TSP studies, albeit one with serious consequences for numerous disciplines.

The 48-States Problem

The challenge of the 1940s was to route a salesman from Washington, D.C., through each of the 48 states in the United States, and back to Washington. Julia Robinson narrowed this down by proposing the salesman visit each of the state capitals, but it does not appear that anyone took the step of writing out a table of travel distances to specify completely the problem, most likely because a solution for such a large instance of the TSP appeared well out of reach.

Dantzig, Fulkerson, and Johnson clearly had a different opinion of the solvability of the challenge, and, without access to a standard set of travel

Figure 4.1
Cities in the 48-States Problem.

distances, went ahead and created their own version of the data. They opted for the quite different selection of cities displayed in figure 4.1; they hit every state, but only twenty of the locations are capitals. Despite this deviation, there is no mystery to their choice. "The reason for picking this particular set was that most of the road distances were easy to get from an atlas."[2] Fair enough, but the selection did give the researchers an immediate head start in the TSP computation: in the standard Rand McNally atlas they consulted, the shortest drive from Washington to Boston passed through seven other cities on the salesman's list. In a bit of a gamble, Dantzig et al. decided to drop these seven northeastern locations. Their reasoning is as follows. If the optimal tour through the remaining cities includes the direct link from Washington to Boston, then the RAND team could solve the original problem by rolling down the window and waving at Baltimore, Wilmington, Philadelphia, Newark, New York, Hartford, and Providence as they drove by. On the other hand, the 42-city tour might reach Washington by some other route, in which case it would have been back to the drawing board.

As you can guess by looking at the map, the optimal tour does indeed use the Washington-Boston link, and thus Dantzig et al. were justified in working with the reduced set of locations. We should point out that things are not so convenient today; using Google Maps, the direct route from Washington to Boston is 451 miles, while adding the remaining seven stops brings the trip to 491 miles. Much of the savings, however, comes from using Interstate 84 through Connecticut and into Massachusetts, and this section of the highway first opened for traffic in 1967.

The data collected from Dantzig's Rand McNally atlas is symmetric, giving distances that do not depend on the direction of travel. (Throughout the chapter we will assume that travel costs are symmetric.[3]) Dantzig et al. adjusted these values by subtracting 10 from each number, then dividing by 17 and rounding the result to the nearest integer. "This particular transformation was chosen to make the d_{ij} of the original table less than 256, which would permit compact storage of the distance table in binary representation; however, no use was made of this."[4] The full table of adjusted distances is contained in their research paper, making precise the problem that had been solved.

Pegs and String

Dantzig et al.'s data distorts somewhat the natural geometry of the problem, but the Euclidean version, where straight-line distances are used, can nonetheless be an effective tool in comparing potential tours. Indeed, the tour-finding approach adopted by the team is based entirely on straight-line approximations.

No hint as to how the USA tour was originally obtained is given in the famous Dantzig et al. paper, but in subsequent lectures Dantzig revealed that a physical device was used. The team constructed a wooden model of the problem, placing pegs at each of the 49 locations, and used a string, tied to a starting city, to wrap around the pegs and trace out a tour. Dantzig described this as a great aid in working with problems by hand; the taut string quickly measures possible routes and identifies likely continuations of subpaths. The model does not provide a solution algorithm in any sense, but with its help Dantzig et al. managed to locate the tour that later proved

Figure 4.2
A peg-and-string tour through Germany. Courtesy of Konrad-Zuse-Zentrum für Informationstechnik Berlin.

to be the optimal route through the 49 cities. Their solution weighed in at 699 units, as measured by the table. Translating this back to atlas distances, the tour covered the United States in 12,345 miles.

Growing Trees and Tours

There is something refreshing about taking a clean sheet of paper, or perhaps a wooden model, and attempting to lay out a good tour. The inclination is to pick a starting point and grow a path, adding one city after another, or, from another point of view, adding one road segment after another. Dantzig et al. relied on intuition to stretch their string from city to city, but simple algorithms can perform the task fairly well.

Nearest Neighbor

If you want to construct a tour, the simplest idea is to always drive to the closest city among those not yet visited. This *nearest-neighbor* algorithm is sensible, although it only rarely finds a shortest-possible solution.

The drawings in figure 4.3 illustrate nearest neighbor in action on the 42-city version of the USA problem, using the distances provided by Dantzig et al. The tour starts in Phoenix and spreads quickly across the southern part of the country. It looks very good for many steps, but when we arrive in the Pacific North West we have no place to go other than to travel all the way back to the East Coast to pick up cities carelessly skipped over during the first pass through the region. This is typical of the algorithm, where we paint ourselves into a corner by not looking ahead when moving from city to city. The final tour in figure 4.3 measures 1,013 units, compared with Dantzig et al.'s optimum of 699 units.

Now, if you are a devil's advocate, you can easily create a TSP instance where nearest neighbor returns a tour that is as bad as you can imagine in comparison to an optimal solution. The point to note is that the algorithm will be forced to take the last leg of the journey, back to the starting city, regardless of its travel cost. So if we increase by 1,000,000 the cost of travel between Montpelier and Phoenix, then poor nearest neighbor will still select the same tour, this time at a total cost of 1,001,013, while the optimal solution remains at 699.

This nasty modification produces a legitimate instance of the TSP, but it does not resemble the types of travel distances we see in road versions of the problem. Indeed, any reasonable instance will satisfy the *triangle inequality*: for any three cities A, B, and C, the cost to travel from A to B plus the cost

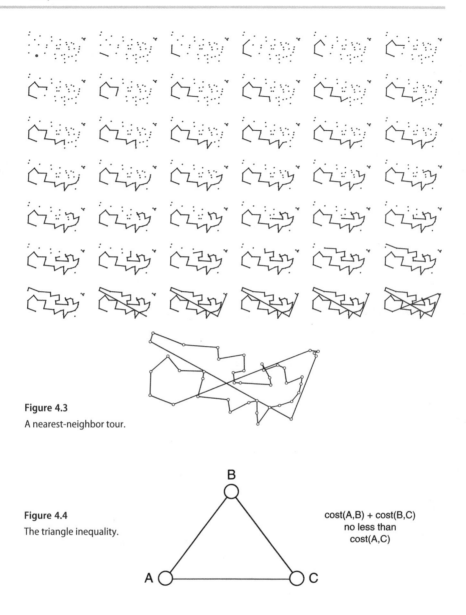

Figure 4.3
A nearest-neighbor tour.

Figure 4.4
The triangle inequality.

cost(A,B) + cost(B,C)
no less than
cost(A,C)

to travel from B to C must not be less than the cost to travel from A directly to C. This condition rules out our nasty case. In fact, it can be shown that with the triangle inequality, and, as usual, symmetric travel costs, nearest neighbor will never do worse than $1 + \log{(n)}/2$ times the cost of an optimal tour for an n-city TSP.[5] So a fifty-city nearest-neighbor tour is guaranteed to be no longer than four times an optimal route, and a million-city tour no worse than eleven times optimal. Perhaps not a great comfort if you

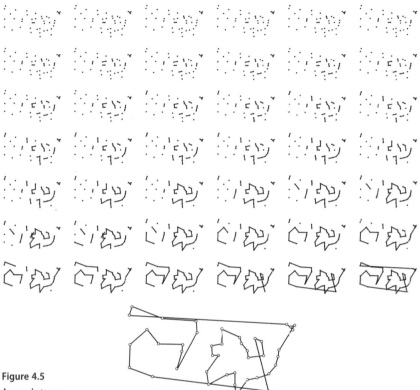

Figure 4.5
A greedy tour.

are depending on the algorithm for your travel plans, but we will see soon methods with better guarantees.

The Greedy Algorithm

Nearest neighbor grows a single path that eventually snakes around and visits every city. The method is a greedy one, extending the path in the shortest possible manner at each step. The name *greedy* is, however, reserved for an alternative algorithm that grows many subpaths simultaneously, adding shortest available road segments wherever they may be found. The operation of the algorithm on the USA problem is illustrated in figure 4.5; the subpaths grow across the map and eventually link up into a tour.

When describing TSP methods such as greedy, it is convenient to adopt graph-theory terminology, with cities being the vertices of the graph and city-to-city road segments the edges. A tour is a Hamiltonian circuit, consisting of a selection of edges corresponding to road segments traveled by the salesman.

The greedy algorithm considers edges in a shortest-first order, adding an edge to the solution only if it joins two subpaths into a longer subpath. The progress of the algorithm looks fantastic early on; the first twenty edges or so in the USA example are very short indeed. The difficulty arises late in the process, when we are forced to accept several very long edges to make the final connections, bringing the tour length up to 995 units.

On large test instances greedy almost always significantly outperforms nearest neighbor. For example, if we drop cities randomly into a square and take straight-line travel distances, then greedy regularly finds tours of length no more than 1.15 times the optimal value, while nearest neighbor produces results in the range of 1.25 times optimal. Unfortunately, this is only an empirical observation. As far as worst-case guarantees go, greedy is known only to do no worse than $1/2 + \log{(n)}/2$ times optimal on instances satisfying the triangle inequality. So just a tiny bit better than the guarantee for nearest neighbor.

Inserting Cities Into a Partial Tour

An immediate question in 1954 was to determine to what extent the Dantzig et al. success relied on the fact that the peg and string model provided an optimal tour, something that could not be counted upon in further studies. In reply, young RAND associate John Robacker jumped in with a series of tests the following summer, solving several 9-city instances with the Dantzig et al. method, starting with random tours. The small examples in his study were not very convincing, but Robacker also described a general tour-finding method that could be automated when attacking large data sets.[6]

> In connection with these experiments, A. W. Boldyreff suggested an approximation procedure, the merit of which lies in its inherent simplicity and in the rapidity with which it may be applied. An application of this approximation method to the 49-city problem of [1] gave a tour of 851 units as compared with the optimal of 699 units, an error of 20%.

The idea is to start with a subtour through a small number of cities and stretch it out, like a rubber band, to enclose one additional city after another.

The Boldyreff/Robacker technique suggests a class of methods called *insertion* algorithms. The algorithms come in different flavors, *cheapest*,

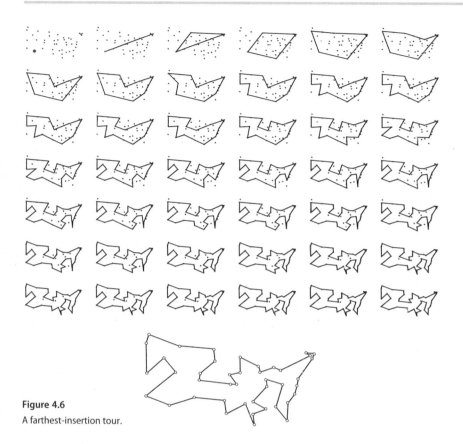

Figure 4.6
A farthest-insertion tour.

nearest, farthest, and *random,* depending on the rule for selecting the next city to add to the growing subtour. In each of these methods the new city is inserted into the spot that causes the smallest increase in the subtour's length.

Robacker described and tested cheapest insertion, where each new city is chosen to be the one that keeps the subtour as short as possible. Nearest insertion chooses the city that gives the shortest distance to any city currently in the subtour; farthest insertion chooses the city that is farthest from the subtour cities; and random insertion selects the next city at random from among those not yet in the subtour.

My favorite among these algorithms is farthest insertion; it obtains a good overall shape for a tour early on, and then completes the details as the last cities are added. The growth process for this variant is illustrated on the USA problem in figure 4.6, starting at Phoenix, expanding out to New Orleans, Minneapolis, and the two Portlands in stage five, and gradually building a tour of length 778.

Cheapest and nearest insertion have both been shown to produce tours no worse than twice the length of optimal solutions when the triangle inequality holds.[7] This is quite nice, but it is curious that farthest insertion comes with only a $\log(n)$ guarantee, even though it is generally the best-performing variant in practice.

Mathematical Trees

Nearest neighbor and greedy typically end up with disappointing tours, despite their beautiful-looking early selections. Greed does not pay in routing the salesman. Surprisingly, a greedy method does produce guaranteed optimal solutions to the related problem of selecting a minimum-cost set of roads to connect a group of cities. Such a minimum-cost structure for the USA data set is displayed in figure 4.7. It has length 591 units, and is thus a good bit shorter than an optimal tour.

My academic great-great-great-great-great-grandfather, Arthur Cayley, studied graphs such as that in figure 4.7. Note that the structure is connected and contains no circuits. Cayley used the wholesome name *trees* for such graphs. His mathematics writing has a nice botanical flavor, referring to vertices as "knots." "In a tree of N knots, selecting any knot at pleasure as a root, the tree may be regarded as springing from this root, and it is then called a root-tree."[8] Rather than springing from a root, we will use the structure to fashion a TSP solution, also guaranteed to be no longer than twice the length of an optimal tour. Trees, by the way, were the subject of the mysterious mathematical problem solved by Matt Damon's character in the film *Good Will Hunting*. The notes drawn by Damon in the scene displayed in figure 4.8 describe Cayley's formula for the number of trees with n vertices, together with several small examples.

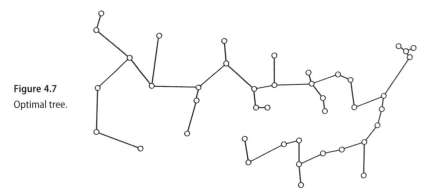

Figure 4.7
Optimal tree.

Figure 4.8
Matt Damon in *Good
Will Hunting*. Copyright
Miramax Films.

You likely have already convinced yourself that an optimal solution to the connection problem will indeed be a tree. The point is that we should never complete a circuit when building a network, since the ends of the final edge are already connected. The greedy algorithm in this case, working in a shortest-first order, includes an edge in the solution only if it is not possible to travel from one of its ends to the other using previously selected edges. The algorithm grows larger and larger connected components until a tree spanning the entire set of cities is produced. It is remarkable, and not too difficult to prove, that this simple method always produces a spanning tree of minimum cost.[9]

A tree is not a tour, but it does give a means to travel from city to city. One way to arrange this is as follows. Whenever we reach a new city, check if it is an end of an unexplored tree edge and, if so, choose such an edge and move along it to reach another city. If, on the other hand, we have already traveled along each of the tree edges meeting the new city, then backtrack until we return to a city that meets unexplored edges. Such a trip is called a *depth-first-search* traversal of the tree. It eventually reaches all cities and backtracks to the start.

The operation of depth-first search is illustrated on a 6-city tree in figure 4.9; the doubled edges are the ones along which we have backtracked. Notice that when the process ends we have traveled along each edge exactly

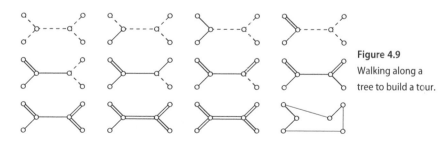

Figure 4.9
Walking along a
tree to build a tour.

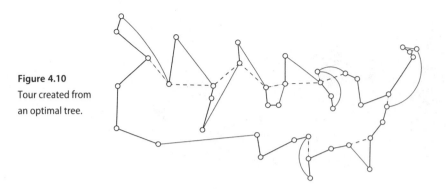

Figure 4.10
Tour created from
an optimal tree.

two times, implying the cost of the trip is twice the cost of the tree. This is good news, since the cost of an optimal tree cannot be more than the cost of an optimal tour. Now, to obtain a tour from the traversal, we simply shortcut over the backtracking steps. These shortcuts are drawn in red in the final tour in figure 4.9.

Applying the algorithm to the USA problem produced the tour of length 823 units displayed in figure 4.10. The depth-first-search traversal in this case started in Phoenix; whenever there was more than one choice for a tree edge to explore, the edge leading to the subtree having the smallest number of cities was taken.

Christofides' Algorithm

Growing a tree to guide a salesman is a nice idea, but to realize its full power we need to step back and view things from the perspective of Leonhard Euler. A depth-first-search traversal of the tree is in fact a Eulerian walk through the graph obtained by duplicating the tree's edges. The duplication step ensures that each vertex of the graph meets an even number of edges, the condition unfortunately violated by the bridges of Königsberg.

Rather than duplicating the tree, we can instead add a set of edges that meets every odd vertex exactly once, where we call a vertex *odd* if it is the end of an odd number of tree edges. The resulting graph has no odd vertices, and therefore admits a Eulerian walk that can be shortcutted into a tour.

To illustrate the idea, figure 4.11 displays the twenty-six odd vertices in the USA tree, and a set of thirteen edges, in red, that meet each of these vertices exactly once. Such a set of edges is called a *perfect matching*, and Jack Edmonds showed how to compute, in polynomial time, a perfect matching of minimum cost. Edmonds's result is a milestone in the field

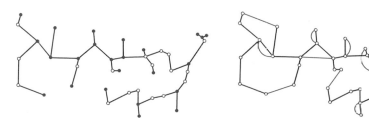

Figure 4.11
A minimum-cost perfect matching of the odd-degree vertices.

Figure 4.12
Nicos Christofides, 1976.

of optimization discussed in chapter 6. For now we note only that this is what the doctor ordered, since, as we argue below, the cost of such an optimal matching can be at most half the cost of an optimal tour. Adding the matching to the tree and shortcutting a Eulerian walk in the resulting graph, we obtain a tour of cost no more than one-and-a-half times that of an optimal solution to the TSP. This is a nice guaranteed performance, and in practice the algorithm typically produces even better solutions. Its operation on the USA problem is displayed in figure 4.13, resulting in a final tour of length 759 units.

Now, to estimate the cost of the optimal matching in general, note first that walking around a TSP tour will take us from odd vertex to odd vertex, with a few even vertices in-between. Shortcutting the even vertices results in a circuit through the odd vertices only, and such a circuit is the union of two perfect matchings, taking every other edge, starting with either the first edge or the second. One of these two matchings must have cost no greater than half the cost of the tour, and Edmonds's optimal matching can only be cheaper still. Voila! This three-step argument is illustrated in figure 4.14,

Figure 4.13
A Christofides tour.

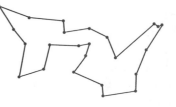

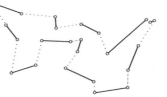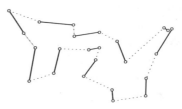

Figure 4.14
Two perfect matchings from an optimal 42-city tour.

where we start with an optimal USA tour, shortcut it to a circuit through the odd vertices, and split the circuit into two matchings.

The full process of combining Euler and Edmonds was strung together by Nicos Christofides in 1976, and it holds a place of honor in the pantheon of the TSP: no polynomial-time algorithm is known to have a better worst-case guarantee than Christofides' method.[10]

New Ideas?

The purity of laying down a tour, piece by piece, is what often attracts people and ideas to the TSP. And it is certainly a good place to gain firsthand experience with the complexity of the problem.

If you want to take a shot at the problem, then improving the performance guarantee of Christofides is a clear target. I must, however, warn you that it may be difficult to beat the factor of one-and-a-half times optimal, as we discuss in chapter 9. On the other hand, it would not be surprising to see new methods that fare well in practical competitions with existing tour-growing algorithms. Alongside the well-known methods we have discussed, TSP fans and researchers have proposed numerous alternatives, including clustering techniques, partitioning methods, spacefilling curves, and more. To date, none of these tour-growing algorithms can beat in practical computation the tour-improvement techniques we treat in the next section, but new ideas could certainly narrow the performance gap.

Alterations While You Wait

A spiffy drawing of a USA tour, displayed in figure 4.15, accompanied Martin Gardner's TSP article in *Discover*, April 1985. The combination of a popular journal and renowned problem solver brought considerable attention to the salesman, but it also stirred up trouble with readers. A close look at the drawing reveals the source of the hubbub: there are obvious shortcuts in the route through the cities!

In a phone conversation with IBM mathematician Ellis Johnson shortly after the article appeared, Gardner described that the tour was in fact obtained from the work of Dantzig et al. The problem did not lie with the tour, but rather with an overzealous editor who went ahead and shifted the locations of the cities over to the 48 state capitals. The *Discover* caption is as follows. "The traveling salesman problem is one of math's most enduring unsolved puzzles. Here's the shortest route for a salesman—or

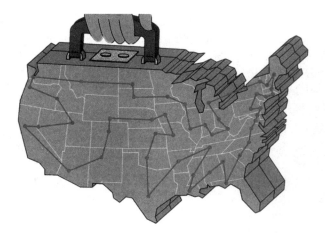

Figure 4.15
United States tour.
Nina Wallace, illustrator,
Discover, April 1985,
page 87.

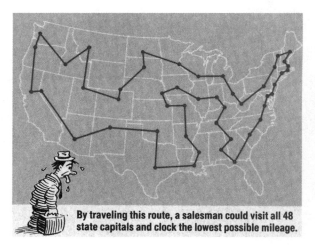

Figure 4.16
Optimal United States
tour. Ron Barrett, illustrator,
Discover, July 1985,
page 16.

By traveling this route, a salesman could visit all 48
state capitals and clock the lowest possible mileage.

salesperson—visiting 48 state capitals." Bad luck. Dantzig et al.'s choice
of convenience in 1954 left Gardner scrambling for a correction to his
publication. This is what led to the phone conversation with Johnson, who
directed Gardner to TSP star Manfred Padberg.

Padberg would certainly have been able to solve the 48-capitals problem, but he presumably could not be reached. In the end it was Shen Lin,
of Bell Labs, who stepped up with a new tour, published in *Discover* four
months after the original. Lin did not have an exact-solution procedure,
but he was a master of tour-improvement methods.

The journal was careful in describing the tour this second time around.
"Is he right? Lin is sure of it. So convinced of his results is he that he's
personally offering a prize of $100 to anyone who can find a route for the
salesman, using his distances between capitals, shorter than 10,628 miles."

The editors sent a table of travel distances to anyone interested in taking up Lin's challenge, but his money was safe. The tour is in fact optimal.

Exchanging Edges

Tour-improvement methods, championed by Lin, do exactly what the name implies. They take as input a tour, search for flaws, and correct them if possible. For example, the spike in the initial *Discover* tour reaching into Tennessee suggests something is wrong with that portion of the route, and the steps outlined in figure 4.18 show how to correct it. We first delete the two edges in the spike and a third edge just to the north, breaking the tour into three segments, one of which is the isolated capital of Tennessee. The segments are rejoined using three new edges, indicated in red. Since the three new edges are together much shorter than the three deleted edges, this *3-opt* move improves the tour.

Figure 4.17
Shen Lin, 1985.
Photograph courtesy
of David Johnson.

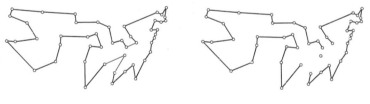

Figure 4.18
Improving the *Discover* tour
with a 3-opt move.

Lin's computation for *Discover* involved an extensive search for tour improvements, including *2-opt* moves, where two edges in a tour are deleted and the tour is reconnected with two shorter edges, 3-opt moves, and more. To explore the ideas he brought to bear on the problem, let's return to the nearest-neighbor tour constructed in our first attempt at the 42-city USA example, a fine candidate for improvement.

Perhaps the oldest theorem concerning the TSP is the fact that for Euclidean instances of the problem an optimal tour will never cross itself. The way to prove this is with a 2-opt move: replacing a crossing pair of edges will always shorten a tour. An obvious move of this type is indicated in figure 4.19. This exchange saves 31 units, bringing the total cost of the tour down to 982 units. And many more such exchanges are available.

By repeatedly making improving 2-opt moves (27 of them altogether), we arrive at the tour of cost 758 displayed in figure 4.20. At this point there exist no further improving moves with just two edges, but this simple process has brought our faulty nearest-neighbor tour to within 8% of the optimal route for the salesman.

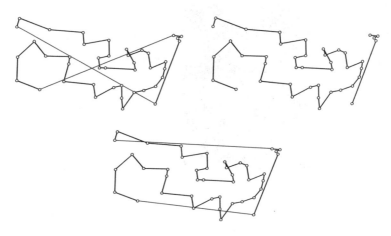

Figure 4.19
An improving 2-opt move for the nearest-neighbor tour.

Figure 4.20
Tour with no further improving
2-opt moves.

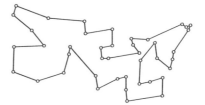

Figure 4.21
Shen Lin and Brian Kernighan, *Bell Labs News*, January 3, 1977. Image courtesy of Brian Kernighan. Reprinted with permission of Alcatel-Lucent USA Inc.

I THINK WE'VE GOT IT, Shen Lin, left, seems to be saying to Brian Kernighan. The MH math and computer experts devised a new, efficient solution to the "Traveling Salesman" problem.

Lin-Kernighan

Bashing on, we could now consider all possible 3-opt moves, checking if any might lead to further improvements. Then 4-opt moves, 5-opt moves, and so on up the line. Success with 3-opt was indeed reported by Lin in the mid-1960s, but the computational burden of searching directly for improving k-opt moves makes the process impractical for k much larger than 2 or 3. Nonetheless, Lin and computer-science pioneer Brian Kernighan accomplished this in a beautifully constructed algorithm.[11] Their work is one of the great achievements of TSP research.

The Lin-Kernighan method is elaborate, but the main idea can be gathered from the sketches in figure 4.22. In the display, the initial tour is laid out as a circle; this makes the process easier to follow, but be aware that the lengths of edges in the sketches are not meant to indicate travel costs.

The search begins by selecting a home city, as well as a tour edge meeting the selected city and a non-tour edge meeting the selected edge's other end. These are indicated by the red city, red edge, and blue edge in the second sketch. Such a triple is considered only if the travel cost of the blue edge is less than the travel cost of the red edge, with the plan being to remove reds and add blues. In the first step we can accomplish such a red-blue exchange by removing also an appropriate tour edge at the far blue end and adding the return segment to the home city, as indicated in the sketch. If this 2-opt move improves the tour, then great, we record how much it saves, but we continue the search in the hope of finding a greater improvement later on.

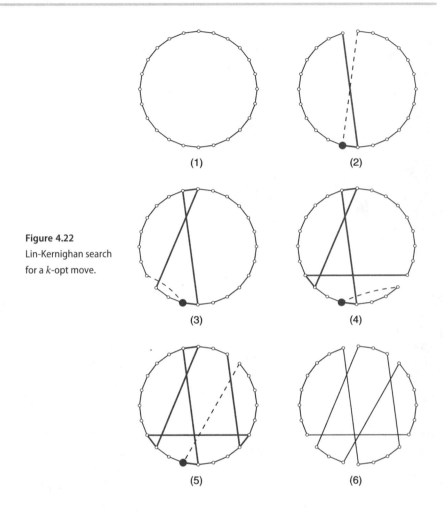

Figure 4.22
Lin-Kernighan search
for a k-opt move.

The next step, illustrated in the third sketch, is to paint the second tour edge red (the one we just tried to delete), and to consider a blue alternative to the direct route home. This extension is explored only if the two blue edges together have cost less than the two red edges. Again in this case, it is possible to come home by removing a tour edge at the far blue end and adding the indicated return segment. We record this potential 3-opt move if it gives the biggest savings thus far.

The search continues to further red-blue pairs, as long as the sum of blue costs is less than the sum of red costs. If we reach the end of the line, where it is no longer possible to add another pair of edges, then we backtrack and explore alternative blue candidates at earlier levels. Eventually we halt the process, either due to time considerations or by running out of edges to consider.

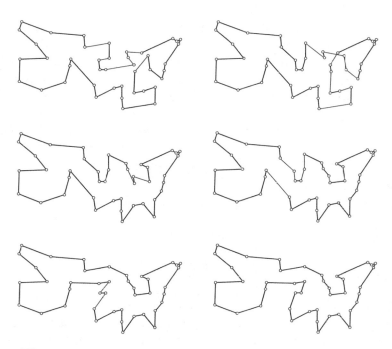

Figure 4.23
Five iterations of Lin-Kernighan.

At the end of the search we take the recorded move yielding the biggest savings, apply it to our tour, and begin again from the newly improved solution. If we failed to find any improving moves, then we return to our starting tour, select a new home city, and attempt another search.

Red-blue, check the home route. Red-blue, check the home route. Sounds easy enough, but there are plenty of devils in the details. Fortunately, together with great computational results, Lin and Kernighan laid out a crystal-clear exposition of their many ideas for implementing and enhancing the search algorithm. With their original paper as a guide, over the past forty years Lin-Kernighan has been engineered to precision, with current implementations capable of producing very good tours to huge TSP instances, having ten million cities and more.

The operation of Lin-Kernighan on the USA data set is illustrated in figure 4.23, starting with the tour obtained from repeated 2-opt moves. The algorithm finds an optimal solution in five iterations; in each step the edges colored red are those that are deleted from the tour.

It should not come as a surprise that the original computer code of Lin and Kernighan also makes short work of the USA example, using random starting tours. "The probability of obtaining optimum solutions in a single trial is close to 1 for small-to-medium problems, say up to the 42-city problem."[12] It is remarkable, however, that their basic method, designed for instances with only several hundred cities or fewer, has served as the cornerstone for the majority of the most successful TSP heuristic methods developed in the past several decades, even as much larger examples have been tackled.

We must note that the great practical performance of k-opt methods is unfortunately not accompanied by great worst-case guarantees. For example, repeatedly making improving 2-opt moves is only guaranteed to produce a solution no worse than $4\sqrt{n}$ times longer than an optimal tour on instances satisfying the triangle inequality.[13] This is the dark side of Lin-Kernighan, but don't be overly concerned when applying the algorithm: this king of methods typically produces very good solutions indeed.

Lin-Kernighan-Helsgaun: LKH

The long reign of Lin-Kernighan in practical computation has been aided by a steady stream of enhancements supplied by the research community. Most of these are tweaks of the original ideas, but computer scientist Keld Helsgaun came along with a bombshell in 1998.

Helsgaun's main contribution was a reworked version of the core search engine, something that had remained basically intact for twenty-five years. Whereas standard Lin-Kernighan can be viewed as a search for a sequence of 2-opt exchanges that taken together result in an improving k-opt move, the new method searches for a sequence of 5-opt exchanges. That is, rather than adopting the step-by-step red-blue search, Helsgaun devised a scheme to consider ten edges at a time, five reds and five blues.

Ten edges. The first thing you should think when you see this is "that's a lot of edges." Indeed, looking at every possibility for five reds and five blues would slow the algorithm to a crawl. To get around this, Helsgaun limits his search to those sets of reds and blues that could potentially be created by a step-by-step red-blue search, if we ignored the condition that the blues must have cost less than the reds at each step. By considering any such *sequential* 5-opt exchange in a stroke, Helsgaun's method can explore improving moves that simply cannot be found by the standard algorithm.

The 5-opt moves, combined with a bag of assorted tricks, allowed LKH to set a new standard in tour finding. "For a typical 100-city problem the optimal solution is found in less than a second, and for a typical 1000-city problem [the] optimum is found in less than a minute."[14] This was an amazing jump in practical performance, in a field of study considered to be quite mature at the time.

Pancake Flipping, Bill Gates, and Big LKH Steps

When news broke that Helsgaun was putting up improved tours for a number of well-known challenge problems, there was plenty of speculation as to how he was able to successfully employ the 5-opt strategy in practical computation. To understand this, I must point out that in the twenty-five years between Lin and Kernighan's research paper and the announcement of LKH, there had been only a small handful of efficient computer codes implementing the standard algorithm. The Lin-Kernighan search method, although well described, is difficult to convert into software that can be run on large data sets.

Lin-Kernighan may be difficult, but LKH would appear to be impossible. Indeed, a great feature of working with a red-blue sequence is that at each step there is only one way to come home. In other words, if we remove two edges from a tour, then there is a unique way to hook up the resulting subpaths to obtain a new tour. A quick calculation shows that LKH, on the other hand, must handle 148 possibilities for joining up the five subpaths involved in a sequential 5-opt move. So 1 versus 148, or difficult versus very, very difficult.

Helsgaun's secret was revealed when he made his entire computer code available to researchers. Going through his files, Dave Applegate and I realized that in fact there was no stealthy method: the code contained a full listing of the 148 cases, independently covering each possibility. Helsgaun had put in a Herculean effort to write a correct and efficient code to implement an extremely complex algorithm.

Helsgaun's code and the performance of LKH were exciting, but it left one wondering if moving up to 6-opt exchanges might be better yet. Dave wrote a small computer code and calculated that sequential 6-opt moves created 1,358 possibilities for reconnecting a tour. That would be daunting enough, but why stop at 6-opt? Well, by the time we get to 9-opt there are a whopping 2,998,656 cases that must be treated. That would be a job indeed.

Not all was lost, however. Dave's code was able to list the reconnection tasks that must be handled, one by one. And an examination of LKH showed a regular pattern in the instructions needed to reconnect the tour. Combining these, we were able to create a computer program that could produce the actual computer code to handle a k-opt move, for any value of k. A computer code building a computer code.

This sounds good, but it resulted in lots of code: 6-opt, 120,228 lines; 7-opt, 1,259,863 lines; and 8-opt, 17,919,296 lines. This was all in the C programming language. Although difficult to compile into a machine workable form, the codes did run and produce interesting results. But 8-opt as a limit was no more satisfying intellectually than 5-opt, and generating the full list for 9-opt was out of the question.

Dave kept up his courage. He had the idea that if we could make the generating code more efficient, then there would be no need to write out the full list of cases. The code-generation method could instead produce the steps needed to handle each case on-the-fly during an execution of a k-opt search. The method would still be limited by the computing time required to execute the search steps, but it potentially permitted the use of much larger moves.

Speeding up these code-generation calculations is closely related to the pancake-flipping problems famously studied by Microsoft's William Gates and TSP expert Christos Papadimitriou, while Gates was an undergraduate student at Harvard University. A flip of a top portion of a stack of pancakes corresponds to a reversal of a subpath in a tour, which is what happens in a 2-opt move. An implementation of a k-opt code generator calls for an algorithm to find a minimum number of flips to rearrange a tour in the order produced by a k-opt move, and this is a variant of the Gates-Papadimitriou work.[15] We managed to get this running, resulting in an efficient on-the-fly search mechanism for sequential k-opt.

Helsgaun incorporated similar ideas into a powerful upgrade to his LKH code, allowing users to specify the size of moves that will be strung together. Demonstrating the reach of the new software, Helsgaun employed 10-opt moves in a computation on a 24,978-city Sweden data in 2003, producing a tour that was shown to be optimal in the following year.

Borrowing from Physics and Biology

Taking a big picture of tour finding, viewing the TSP as just one example of a general search problem, proves to be useful both in finding good tours and in devising multipurpose techniques. The idea is to produce *metaheuristics*, that is, heuristic methods for the design of heuristic methods. The general nature of this work has brought in researchers from fields of science to join in the hunt for good tours.

Local Search and Hill Climbing

A useful analog in this arena is to think of tours as lying on a landscape, with the elevation of each tour corresponding to its quality. The type of picture to have in mind is one like the Gasherbrum group of mountains displayed in figure 4.24: good tours correspond to peaks of the mountains,

Figure 4.24
Gasherbrum group. Image by Florian Ederer.

with the optimal tour lying on top of mighty Gasherbrum II. A heuristic algorithm can be viewed as moving through the landscape in search of high land.

For this picture to make sense there should be a notion of when two tours are located near to one another. This is typically handled by creating *neighborhoods* around each tour. For example, two tours can be defined to be neighbors if one can be reached from the other via a 2-opt exchange, or via an exchange found by Lin-Kernighan. Large neighborhoods are useful for navigating around a landscape, but they should be constructed so that algorithms can view and evaluate neighbors.

Tour-improvement methods, such as repeatedly making improving 2-opt moves, are often called *hill-climbing* algorithms, since they can be viewed as walking up a sequence of neighboring tours, always moving to higher ground. At each step we do a *local search* for a nearby higher point. If we are thorough in our search, then the algorithm will terminate at a peak, or at least a plateau, since at this point all local moves will be either downhill or flat. A full run of the algorithm begins at the point corresponding to the starting tour and then scoots up a slope to reach a local peak.

Note that the choice of a starting tour can determine the fate of a hill-climbing approach: if the starting tour lands midway up a small hill, then the algorithm will be limited to reaching the modest-quality tour associated with the hill's peak. For this reason Lin and Kernighan proposed to carry out repeated runs of their algorithm from random starting solutions. The idea is to throw darts into the landscape. If we toss enough darts, then there is a decent chance of hitting a slope leading to a peak of good height.

Random tours provide a nice distribution of darts, but they have the disadvantage of typically starting far down in valleys, due to their poor tour quality. For a large TSP instance it can take a long time to walk from a valley to a peak. A compromise approach is to use nearest-neighbor darts, gaining randomization from the selection of the starting city.

Simulated Annealing

In *simulated annealing* heuristics, the hill-climbing strategy is relaxed to allow the algorithm to accept with a certain probability a neighbor that is worse than the current solution. At the start the probability of acceptance is high, but it is gradually decreased as the run progresses. The idea is to allow the algorithm to jump over to a better hill before switching to a steady climb.

The paper of Scott Kirkpatrick, Daniel Gelatt, and Mario Vecchi that introduced the simulated-annealing paradigm studied the TSP, reporting on heuristic tours for a 400-city problem.[16] The authors of the paper write that the motivation for the method comes from a connection with statistical mechanics, where annealing is the process of heating a material and then allowing it to slowly cool to refine its structure.

For the salesman, the achievements of the paradigm have to date been rather modest. But as a general search tool simulated annealing has been a spectacular success. Google Scholar lists over 18,700 citations to the original research paper, an almost unheard of number.

Chained Local Optimization

The greatest impact of simulated annealing on current tour-finding methods is perhaps not the technique itself, but rather the fact that it brought the thinking of physics into the TSP arena. Indeed, it was a second major contribution from physicists that first pushed computational results beyond the limits of repeated Lin-Kernighan.

In the late 1980s, Olivier Martin, Steve Otto, and Edward Felten, from the physics department at Caltech, proposed an alternative to the dart-throwing strategy. The idea is to take advantage of the fact that a strong local-search algorithm, such as Lin-Kernighan, will typically take us up into the high-elevation region of the tour landscape. Rather than starting a second run of the algorithm from a random location, Martin et al. suggest we first look around our current peak to see if there might not be a way to jump over a few local barriers to reach a new slope to take us to an even better location.

The specific proposal is to *kick* the Lin-Kernighan solution to obtain a new starting tour, rather than throwing a dart. The overall process repeats

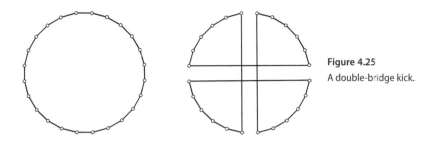

Figure 4.25
A double-bridge kick.

this many times, replacing the solution whenever we reach a better tour. For the method to work, a kick must take the solution out of its neighborhood, so it should be a modification that Lin-Kernighan cannot easily undo. Martin et al. found that a random 4-opt exchange of the type indicated in figure 4.25 does the job nicely.

The resulting algorithm is dubbed *Chained Lin-Kernighan* and its performance is outstanding. The stars were aligned for this idea. First, Martin et al.'s intuition was correct: visiting the nearby region via the kicking mechanism is a better way to sample the peaks in the landscape; we use Lin-Kernighan itself to guide us to the highest elevations. Second, the reapplication of Lin-Kernighan to a kicked tour runs much more quickly than an application to a random tour. This is simply due to the fact that much of a kicked tour remains in good shape, so the algorithm does not need many iterations to reach a locally optimal result.

For most of the 1990s, implementations of Chained Lin-Kernighan ruled the world of tour finding. The version included in the Concorde code routinely finds, in one or two seconds, solutions within 1% of the cost of optimal tours for instances with up to 100,000 cities. For even better solutions, one can turn to LKH, but Chained Lin-Kernighan remains dominant on very large data sets. For example, the plot in figure 4.26 shows the results of a run on a 25,000,000-city Euclidean instance, with city locations having integer coordinates drawn at random from a 25,000,000 × 25,000,000 square. In eight days, on a computer from the year 2000, a tour that is approximately 0.3% greater than optimal was found.[17]

Genetic Algorithms

An alternative to the landscape view is to consider a salesman's route as a living organism, mutating and evolving over time. This way of thinking is taken up in a class of methods known as *genetic algorithms*, inspired by John Holland's landmark book *Adaptation in Natural and Artificial*

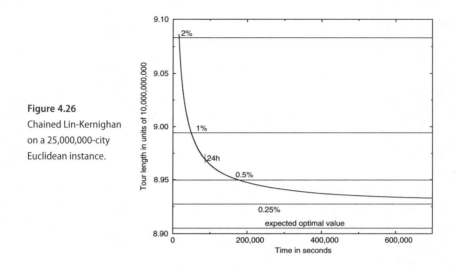

Figure 4.26
Chained Lin-Kernighan
on a 25,000,000-city
Euclidean instance.

Systems published in 1975.[18] Holland did not treat the TSP, but his ideas quickly made their way into the tour-finding literature.

A general outline of a genetic algorithm, as applied to the salesman, is the following. We begin by generating a starting population of tours, say by repeatedly applying nearest neighbor with random starting cities. In a general step, we select some pairs of members of the population and *mate* them to produce *child* tours.[19] A new population of tours is then selected from the old population and the children. The process is repeated a large number of times and the best tour in the population is chosen as the winner.

The spirit of genetic algorithms is to mimic evolutionary processes found in nature. The analogy is fun, but keep in mind that merely adopting the language of Darwin does not imply we end up with a good tour. Indeed, early genetic algorithms for the TSP were not especially successful, even while restricted to very small instances of the problem. But the idea of maintaining a population of tours has considerable merit and the general approach can be crafted into very strong heuristics, particularly in combination with local-search procedures.

The genetic-algorithm outline leaves plenty of freedom for selecting methods to evolve a tour population. Besides the mating process, we also get to choose a *fitness measure* for selecting the next population. Some cool ideas have been developed for such measures, seeking to balance the quality of solutions with the need for a diverse population.

For mating itself, early schemes attempted to find subpaths in one parent tour that could be substituted for subpaths in the other parent. This

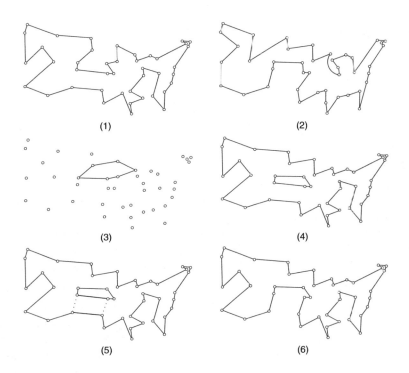

Figure 4.27
Mating two tours.

was rather restrictive, particularly in larger instances. A more successful approach is to create a new tour by choosing a subpath in parent A and extending it to a tour using, when possible, edges of parent B or edges of parent A, with preference given to the edges of B. Another mating technique, known as *edge-assembly crossover (EAX)*, is illustrated on a pair of USA tours in figure 4.27. To combine the blue and red solutions in the example, we form the graph consisting of the union of their edges and select a circuit, displayed in the third sketch in figure 4.27, that alternates between blue and red. We then delete from the blue tour each of the circuit's blue edges and add to the blue tour each of the circuit's red edges, as illustrated in the fourth sketch. The process creates subtours that are combined into a tour via a 2-opt move, displayed in the fifth and sixth sketches in the figure.

The EAX mating scheme was adopted by Yuichi Nagata in one of the most successful tour-finding procedures proposed to date.[20] His implementation relies on a very fast implementation of EAX, allowing the algorithm to proceed through many generations of tours. Among Nagata's achievements is his discovery of the best-known tour for the 100,000-city Mona Lisa TSP.

Ant Colonies

At some point in your life you likely had the misfortune of losing food to a group of hungry ants. Typically the pests arrive in your home or garden via a long thin train of individuals, constantly moving back and forth in a nearly straight line. A single ant moves haphazardly, but the entire group, communicating via pheromone trails, finds an efficient route. This collective behavior is the inspiration for a class of TSP heuristics known as *ant-colony optimization* (ACO).

The leader of ACO research is Belgium's Marco Dorigo, who developed the ideas in his 1992 Ph.D. thesis.[21] His algorithms work with a small army of ant agents moving along the edges of a graph. Each agent traces out a tour, selecting at each new vertex an edge chosen among those leading to vertices not yet visited. The key to the process is the selection rule, which makes use of a *pheromone value* associated with each edge; if an edge has a high pheromone value then it has a high probability of being selected. After the agents have all completed tours, the pheromone values are adjusted using a rule that adds values proportional to the lengths of the computed tours; edges in good tours get their values increased more than those in poor tours.

The approach is both intuitive and appealing, but thus far ACO has not proved to be competitive with Lin-Kernighan-based methods. In recent years, however, the paradigm has been applied effectively to problems in

Figure 4.28
Ants working on the TSP.
Image by Günter Wallner.
Originally appeared in the
book *Bilder der Mathematik*
by Georg Glaeser and Konrad
Polthier.

other areas, such as scheduling, graph coloring, classification, and protein folding. The active research topic is a good example of how a focus on the salesman can lead researchers to interesting general-purpose methods for attacking optimization problems arising in diverse applications.

And Many More

We have touched on only the best-performing applications of metaheuristic ideas for the TSP. Other schemes include neural networks, tabu search, and honeybee models, to name just a few. If you have a general search mechanism in mind, the TSP is a great place to develop, polish, test, and compare your strategy, even if your planned domain of application is far away from the humble routing of a salesman.

The DIMACS Challenge

The breadth of activity in tour finding is a strength of the area, but it has in the past led to misunderstandings concerning the state of the art. Indeed, in the 1980s research papers appeared in premier scientific journals, such as *Nature*, describing computations on TSP instances having 30 or 50 cities. The reported results were typically weak approximations, at a time when Lin-Kernighan could reliably deliver optimal solutions in a blink of an eye, and Martin Grötschel and Manfred Padberg were tackling instances with hundreds of cities via exact methods.

This difficulty was addressed by two important events in the following decade. The first of these was the TSP 90 conference held at Rice University's Center for Research in Parallel Computing. The organizers brought together exact-solution experts such as Grötschel and Padberg, together with tour-finding teams from around the world. An important outgrowth of the meeting was the establishment of the TSPLIB collection of test problems by Gerhard Reinelt from the University of Heidelberg. Reinelt's library was published in 1991, containing over 100 challenge instances of the TSP gathered from academic and industrial sources. The TSPLIB collection provides a common test bed for researchers around the world and across academic disciplines.[22]

The second event was the DIMACS TSP Challenge, led by David S. Johnson of AT&T Research. DIMACS is the short name for the Center for Discrete Mathematics and Theoretical Computer Science, housed at Rutgers University. In the 1990s DIMACS ran a series of implementation challenges, the best known of which is the TSP Challenge.[23]

Figure 4.29
Left: Martin Grötschel, Gerhard Reinelt, and Manfred Padberg.
Right: Robert Tarjan, Dorothy Johnson, Al Aho, and David Johnson.

One goal of this Challenge is to create a reproducible picture of the state of the art in the area of TSP heuristics (their effectiveness, their robustness, their scalability, etc.), so that future algorithm designers can quickly tell on their own how their approaches compare with already existing TSP heuristics.

DIMACS made a call to the world's tour finders, and the world responded with 130 different algorithms and implementations. A great outcome of the challenge is a Web site that allows for direct comparisons between methods. The results are also gathered together in a very nice survey paper by Johnson and co-organizer Lyle McGeogh.[24]

Johnson's efforts in organizing the challenge, as well as his own detailed computational studies of tour-finding methods, have been a great force in shaping the current area of algorithm engineering. In 2010 he received the Knuth Prize from the Association for Computing Machinery, cited for his contributions to the theoretical and experimental analysis of algorithms. A well-deserved recognition for one of the world's leaders in TSP research.

Tour Champions

Heuristic methods must strike a balance between running time and tour quality. At the highest end of the scale we are willing to spend enormous amounts of time to deliver the best solution that is practically possible. This is Formula One racing, with participants in a no-holds-barred contest to push down the lengths of best-known tours through challenge data sets.

Figure 4.30
Left: Keld Helsgaun.
Right: Yuichi Nagata.

The world champions in this area are without a doubt Keld Helsgaun of Denmark and Yuichi Nagata of Japan. Helsgaun's LKH code has been the gold standard in tour finding since its introduction in 1998, and he has continued to extend and improve his algorithm with many new ideas. Helsgaun is the current holder of the best-known tour in the World TSP challenge, he provided the optimal tour for the record 85,900-city TSP, and his name peppers the leader board for the *VLSI Test Collection*.[25] Not to be outdone, Nagata's implementation of a genetic algorithm for the TSP has produced the best-known tour in the Mona Lisa TSP challenge as well as record solutions for the two largest examples in the *National TSP Collection*.[26] If you want a good solution to a large problem, these are the people to call.

5: Linear Programming

The development of linear programming is—in my opinion—the most important contribution of the mathematics of the 20th century to the solution of practical problems arising in industry and commerce.
—Martin Grötschel, 2006.[1]

S electing the best tour through a set of points and knowing it is the best is the full challenge of the TSP. Users of a brute-force algorithm that sorts through all permutations can be certain they have met the challenge, but such an approach lacks both subtlety and, as we know, practical efficiency. What is needed is a means to guarantee the quality of a tour, short of inspecting each permutation individually. In this context, the tool of choice is *linear programming*, an amazingly effective method for combining a large number of simple rules, satisfied by all tours, to obtain a single rule of the form "no tour through this point set can be shorter than X." The number X gives an immediate quality measure: if we can also produce a tour of length X then we can be sure that it is optimal.

Sounds like magic, but linear programming is indeed the method adopted in Concorde and in all of the most successful exact TSP approaches proposed to date. Moreover, its application to problems beyond the TSP has made it one of the great success stories of modern mathematics.

General-Purpose Model

The tale of linear programming has a nice start, with a young George Dantzig arriving late for a class given by Jerzy Neyman at the University of California at Berkeley in 1939. The first-year graduate student hurriedly copied down two problems he found written on the board and turned in solutions several days later. "To make a long story short, the problems on the blackboard that I had solved thinking they were homework were in fact two famous unsolved problems in statistics."[2] Not a bad week's work. The solutions ended up being the main content of Dantzig's Ph.D. thesis.

After Berkeley, Dantzig spent the years of World War II studying programming problems for the United States Air Force. In this military context, *program* is not a reference to a set of computer instructions, but rather to "proposed schedules of training, logistical supply and deployment of combat units."[3] Dantzig became an expert at delivering such programs, using desk calculators to crunch numbers provided by various reporting systems.

At the conclusion of the war effort, Dantzig was offered an attractive position to keep him at the Pentagon. Together with a nice salary, the offer came with a specific research target: his colleagues Dal Hitchcock and Marshall Wood set Dantzig the goal of mechanizing the military's program-planning process. Not one to hide from a challenge, Dantzig took the bull by the horns and devised the far-reaching theory that became known as linear programming, or *LP* for short.

Linear Programming

Dantzig's LP research was strongly influenced by the work of Wassily Leontief, who in the 1930s developed an economic model specifying a balance between inputs and outputs of production. Dantzig extended this idea with a general notion of constraining choices in economic activities.

Figure 5.1
George Dantzig. Photograph
courtesy of Mukund Thapa.

Three elements are key In Dantzig's model. First, rather than dealing with balance equations only, his LP constraints may also include *inequalities*, expressing that one quantity must be at least as large as another. A main use of this feature is to state that the amount of some item must be at least zero, as Dantzig explained with the Mad Hatter's help.[4]

"Take some more tea," the March Hare said to Alice, very earnestly.

"I've had nothing yet," Alice replied in an offended tone, "so I can't take more."

"You mean you can't take *less*," said the Hatter: "it's very easy to take *more* than nothing."

Precisely so. Some items, such as the quantity of tea consumed, only make sense if they are nonnegative.[5] Thus, if a variable T represents the quantity of tea taken by Alice, then we have as a constraint that T must be at least 0. The shorthand for writing such a constraint is $T \geq 0$, where the symbol "$\geq$" stands for greater-than-or-equal-to.

The second key idea is a restriction to *linear* constraints. William Safire, in his "On Language" column, weighed in on the use of the word. "Linear thinking is generally a put-down, synonymous with 'unimaginative' or 'too logical,' but linear programming tries to deal with the way all parts of a system interact with all the other parts over time."[6] In Dantzig's logic, activities are assumed to consume resources in proportion to their levels. If Alice takes two lumps of sugar in one cup of tea, then we assume she would like four lumps of sugar in two cups of tea: doubling the level requires double the resources. The expression in this case is $S = 2T$, that is, $S - 2T = 0$, where S is the quantity of sugar. You will recognize $S - 2T = 0$ as an equation of a line, and hence the name "linear."

In a general constraint, a collection of variables representing levels of activities can be combined by taking multiples of the variables and adding them up. So, if we have variables A through Z, possible constraints in an LP model include

$$A + B + C + D \geq 100,$$

$$2E + 8G - H = 50,$$

and

$$1.2Y - 3.1X + 40Z \geq 0.$$

It is not permitted to multiply variables together, such as $XY \geq 0$, or to take square roots or other fancy constructions you might be tempted to include. This is a real limitation, but it is this linear restriction that ties the model together from a computational perspective.

The third key element of linear programming is the inclusion of an explicit *objective* to provide a means for ranking candidate solutions. Dantzig viewed this as one of his great practical accomplishments, forcing officers and managers to be precise in expressing what they want to achieve. In specifying this objective, the value of an activity is assumed to be proportional to its level. We thus require the objective to be a linear expression of the variables in the model, and this expression is maximized, or minimized, over all allowable assignments of levels to the activities. An *optimal solution* to an LP model is an assignment such that the objective is as large, or small, as possible.

This is the full framework, but to meet the challenge of Hitchcock and Wood, Dantzig also needed a means to deliver optimal solutions to modeled problems. His answer in this case is a computational tool called the *simplex algorithm*. You feed the algorithm LP data and it pops out an optimal solution. The algorithm is too important, both in general applications and for the TSP in particular, to cover in a short description, so let's postpone a discussion until the next section.

Widget Incorporated

Given the large number of definitions in the past couple of pages, it may be useful to consider a small example to demonstrate how everything fits together. To keep things simple, suppose we manufacture three types of widgets, the items so loved by economists. Call them widget A, widget B, and widget C, with their names associated to variables specifying the quantities of each. Widget production requires the input of two raw materials, say nickel and steel, and we have in stock 100 pounds of the first and 200 pounds of the second. To manufacture widget A we need 3 pounds of nickel and 4 pounds of steel; widget B requires 3 pounds of nickel and 2 of steel; widget C requires 1 pound of nickel and 8 of steel.

Total profit earned through widget production and sale is $10A + 5B + 15C$, that is, model A brings in \$10 per unit, model B brings in \$5 per unit, and model C brings in \$15 per unit. The problem is to find levels of production that earn as much money as possible, without exceeding our stocks of raw materials. For example, focusing on the most profitable model, we could produce 25 units of widget C before we run out of steel. This plan brings in \$375, but linear programming tells us how to do better. The model is the following, where the symbol "$\leq$" means less-than-or-equal-to:

$$\text{maximize } 10A + 5B + 15C$$

$$\text{subject to}$$

$$3A + 3B + 1C \leq 100 \quad \text{(nickel constraint)}$$

$$4A + 2B + 8C \leq 200 \quad \text{(steel constraint)}$$

$$A \geq 0, \, B \geq 0, \, C \geq 0.$$

The first constraint ensures we have a sufficient supply of nickel to meet our production plan, and the second constraint ensures we have a sufficient supply of steel.

An optimal solution to the LP model is to earn \$450 by producing 30 units of widget A, none of widget B, and 10 units of widget C. Easy enough, even without the help of the simplex algorithm. But an actual production problem can include hundreds of thousands of variables and constraints. Not something you would want to try to solve in your head.

A Linear World

News of the general linear-programming model, and the simplex algorithm for its solution, was delivered by Dantzig in 1948 at a meeting held at the University of Wisconsin. The event was a defining moment for Dantzig, who has described often its proceedings. Like many good stories, repeated telling may have shifted a few details over the years, but all versions capture the spirit of a nervous rising star facing a large and distinguished group of mathematicians and economists.[7] During the discussion following Dantzig's lecture, Harold Hotelling, great in both academic stature and physical size, rose from his seat, stated simply, "But we all know the world is nonlinear," and sat down. Dantzig was lost for a reply to such a sweeping criticism.[8]

> Suddenly another hand in the audience was raised. It was von Neumann. "Mr. Chairman, Mr. Chairman," he said, "if the speaker does not mind, I would like to reply for him." Naturally I agreed. von Neumann said: "The speaker titled his talk 'linear programming' and carefully stated his axioms. If you have an application that satisfies the axioms, well use it. If it does not, then don't."

Fortunately for the world, many of its complexities can in fact be described in sufficient detail by linear models. The episode with Dantzig, Hotelling, and John von Neumann is summed up nicely by a cartoon Dantzig's Stanford colleagues reported as hanging outside his office.[9] It featured the *Peanuts* character Linus in his traditional pose, sucking his thumb and holding a blanket. The caption read, "Happiness is assuming the world is linear."

Applications

George Dantzig's classic book *Linear Programming and Extensions* begins with the following statement. "The final test of a theory is its capacity to solve the problems which originated it."[10] This is a bold way to open a 600-page volume, but sixty years of experience has shown repeatedly that his theory of linear programming and the accompanying simplex algorithm pass the test beyond all expectations.

The scope of the use of linear programming in industry is breathtaking, covering pretty much any sector you can name. Although it is difficult to quantify, it is clear that planning via linear programming saves enormous amounts of the world's natural resources every day. It terms of money, a *New York Times* article by Gina Kolata states, "Solving linear programming problems for industry is a multibillion-dollar-a-year business."[11] Take that, Professor Hotelling.

Readers interested in learning the art of capturing problems with linear constraints can find what they are looking for in Paul Williams's excellent book *Model Building in Mathematical Programming*.[12] Williams's examples include food manufacturing, refinery optimization, farm planning, mining, airline pricing, power generation, and on and on.

The Simplex Algorithm

How often do you see mathematics described on the front page of the *New York Times*? The proof of Fermat's Last Theorem made the cut, but the Four-Color Theorem missed out. It is with some pride that linear-programming researchers point to two cover stories.

A surprise discovery by an obscure Soviet mathematician has rocked the world of mathematics and computer analysis, and experts have begun exploring its practical applications.
—Malcolm W. Browne, *New York Times*, November 7, 1979.

A 28-year-old mathematician at A.T.&T. Bell Laboratories has made a startling theoretical breakthrough in the solving of systems of equations that often grow too vast and complex for the most powerful computers.
—James Gleick, *New York Times*, November 19, 1984.

Each announces a new polynomial-time algorithm to solve LP problems, the first by Leonid Khachiyan of the former Soviet Union and the second by Narendra Karmarkar of Bell Labs.

Both front-page articles suggest, somewhat aggressively, the imminent demise of the simplex algorithm for the practical solution of large-scale problems. But Dantzig's method was not to be dethroned so easily. New computer implementations in the late 1980s, including Bob Bixby's *CPLEX* code at Rice University and John Forrest's *OSL* code at IBM, showed the algorithm had plenty of punch left. Indeed, in the year 2000 it was named one of "The Top Ten Algorithms of the Century" and it remains the workhorse of the field of mathematical optimization.[13]

Pivoting to a Solution

A generation of applied mathematicians and engineers have learned the nitty-gritty of the simplex algorithm. There are a number of fine texts offering a wide range of schemes for presenting the step-by-step procedure, but none better than Vašek Chvátal's beautifully written book *Linear Programming*.[14] We follow his presentation by means of an example, supplied by our widget manufacturer.

The variables A, B, and C in the widget LP indicate the levels of production for the three models offered for sale. These values are enough to specify completely the problem, but in a report to management we would probably like to include as additional information the total profit, say Z, and the remaining stocks of nickel and steel, say N and S. The new variables Z, N, and S are defined by equations

$$Z = 0 + 10A + 5B + 15C$$

$$N = 100 - 3A - 3B - C$$

$$S = 200 - 4A - 2B - 8C$$

that serve as a *dictionary* for looking up their values.

A natural, although disappointing, assignment associated with the dictionary is to set each of A, B, and C to zero, and read off the profit $Z = 0$, the nickel stock $N = 100$, and the steel stock $S = 200$. Disappointing, but easy to improve. For example, since we have the term $10A$ in the profit equation, if A is increased from its current zero level then additional money rolls into the firm. But to what level can we increase A while satisfying the model's constraints? Keeping $B = 0$ and $C = 0$ will maintain the three nonnegativity constraints, but we must also ensure that we do not overuse the stocks of raw materials, that is, we must have $N \geq 0$ and $S \geq 0$. The first of these two constraints implies $100 - 3A \geq 0$, or, in other words, we can increase A to at most $33\frac{1}{3}$ before we run out of nickel. Similarly, the inequality for S implies we can increase A to at most 50 before we exhaust our stock of steel. In this case nickel is the binding constraint, so let's go ahead and set

$$A = 33\frac{1}{3}, \ B = 0, \ C = 0,$$

bringing in a profit of $\$333\frac{1}{3}$.[15]

Now, what about increasing B or C from their current zero levels? It is not so easy to see what will happen in this case, since we will be forced to also decrease the level of widget A to free up the depleted stock of nickel. To make the dependence on A clear, we can rewrite things, moving A over to the left-hand side of the dictionary. We do this by moving N to the right-hand side, since it is now at the zero level.

The dictionary's equation for N is

$$N = 100 - 3A - 3B - C,$$

or, changing the sides of A and N and dividing by three,

$$A = 33\frac{1}{3} - \frac{1}{3}N - B - \frac{1}{3}C.$$

We will use this definition of A in a new dictionary, but we need first to substitute the expression for the appearance of A in the definitions of the profit Z and the stock variable S. Doing so yields the system of equations

$$Z = 333\frac{1}{3} - 3\frac{1}{3}N - 5B + 11\frac{2}{3}C$$

$$A = 33\frac{1}{3} - \frac{1}{3}N - B - \frac{1}{3}C$$

$$S = 66\frac{2}{3} + 1\frac{1}{3}N + 2B - 6\frac{2}{3}C.$$

This operation of switching the roles of A and N is known as *pivoting*; A is called the *incoming variable* and N the *leaving variable*. Note that although we have new equations for Z and S, the variables still measure our total profit and our stock of steel, respectively. Indeed, since we used an equation satisfied by all assignments to rewrite the expressions, the variables maintain their old meanings.

With this new dictionary, the natural solution is to set N, B, and C to zero, and to read off the values

$$Z = 333\frac{1}{3}, \quad A = 33\frac{1}{3}, \quad S = 66\frac{2}{3}.$$

A profit of $\$333\frac{1}{3}$ is more than a profit of zero, but we can do better. Examining the rewritten profit equation, we see negative terms for N and B, implying that profit will decrease if we attempt to raise either of their levels. Fortunately, the term for C is positive, so we should attempt to increase its value with another pivot operation.

Let's go through the steps to pivot on C as the incoming variable. First, keeping N and B at zero levels, the new dictionary tells us

$$A = 33\frac{1}{3} - \frac{1}{3}C \geq 0$$

and thus C can be increased to at most 100. Similarly,

$$S = 66\frac{2}{3} - 6\frac{2}{3}C \geq 0,$$

implying C can be increased to at most 10. Now 10 is smaller than 100, so S is the leaving variable in the pivot operation.

The next step is to rewrite the dictionary, exchanging the roles of C and S. Skipping over the arithmetic, we arrive at the new set of equations

$$Z = 450 - N - \frac{3}{2}B - \frac{7}{4}S$$

$$A = 30 - \frac{2}{5}N - \frac{11}{10}B - \frac{1}{20}S$$

$$C = 10 - \frac{1}{5}N - \frac{3}{10}B - \frac{3}{20}S.$$

The natural solution corresponding to this dictionary is

$$N = B = S = 0, \quad Z = 450, \quad A = 30, \quad C = 10.$$

This is the assignment we claimed was optimal in the previous section. And now you need not take my word for it: the profit equation $Z = 450 - N - \frac{3}{2}B - \frac{7}{4}S$ is proof that the widget producer can never earn more than $\$450$. Indeed, N, B, and S must all be nonnegative in any assignment, thus our profit is $\$450$ minus a nonnegative number. And that is that.

The simplex algorithm thus proceeds from dictionary to dictionary via pivot operations, attempting to increase the objective at each step. The process ends when the objective equation proves that the current natural solution is optimal. Easy enough, but we are sweeping a few details under the rug: How do we in general find a starting dictionary? How can we be sure the method eventually terminates? These can both be handled, but allow me to point you to Chvátal's book for a full analysis.

If you would like to work out additional examples, but don't find appealing the idea of carrying out so much arithmetic by hand, then pop over to the Web page for Bob Vanderbei's *Simplex Pivot Tool*.[16] The small tool allows you to set up your own LP model as a dictionary and to carry out pivots by selecting the incoming and leaving variables. The Web page will even warn you if you make a mistake, that is, if your choice of a leaving variable is not one corresponding to a binding constraint.

A Polynomial-time Pivot Rule?

Dantzig delivered an algorithm, but why should we think it is practical for solving a problem of any respectable size? A fair question.

In 2009 Richard Karp presented the opening lecture at a meeting held in Atlanta to celebrate the fiftieth anniversary of the Foundations of Computer Science conferences. His lecture was titled "What makes an algorithm great?" and first on his list of examples was the simplex algorithm, cited for its practical efficiency and its wide use. Karp had to point out, however, that the method is not known to be great from the complexity point of view. Indeed, the simplex algorithm comes without a guarantee of polynomial running time.

The issue is that although the number of possible dictionaries is finite, there are exponentially many of them. It is not known if there exists a plan for choosing the incoming and leaving variables at each step in such a way that we are sure to reach an optimal dictionary after only a polynomial number of pivots. In fact, it is known that several natural pivot rules are not polynomial, that is, specific pathological examples have been constructed that cause the simplex algorithm, equipped with the particular rule, to run through an exponentially long sequence of pivots.

So why did Dantzig have so much faith in the algorithm? The answer is that he did not, at least not initially.[17]

> At first I thought that the method might be efficient but not nec-
> essarily practical. For a big problem there could be many combina-
> tions (corner points)—perhaps as many as the stars in the heavens.
> It might require a million steps to solve it. That might be considered
> efficient, since this number is small relative to the number of

combinations involved, but hardly practical. So I continued to look for a better alternative algorithm.

It was only after his Pentagon colleagues succeeded in solving problem after problem that the algorithm eventually won Dantzig's endorsement.

This remains the situation today. The simplex algorithm is one of the most widely used tools in mathematics, but we don't know for sure that it will continue to solve all of the instances that arise in practical models. This is the real content of the front-page stories in the *New York Times*. Fame, and perhaps practical fortune, awaits anyone who can discover a pivot rule that is guaranteed to reach an optimal solution in a polynomial number of steps.

A Million Times Faster

The short abstract describing Dantzig's lecture at the Econometrica Society's 1948 meeting concludes with the following line. "It is proposed that computational techniques such as those developed by J. von Neumann and by the author be used in connection with large scale digital computers to implement the solution of programming problems."[18] The dawn of digital computing was an exciting time for applied mathematicians, and Dantzig was pushing the simplex algorithm to the head of the line for a possible computer implementation.

The first large-scale test of the simplex algorithm was, however, accomplished without the use of a computer. The computation was carried out at the National Bureau of Standards in 1947, involving a team working away with hand-operated desk calculators for a total of 120 man/woman days.[19] The test instance modeled a problem of selecting an adequate diet at minimum cost, and contained 9 constraints and 77 nonnegative variables. That is large, although a far cry from the hundreds of thousands of constraints and variables that go into current industrial LP models. Fortunately, in the years between this diet-problem computation and the release of modern LP software, plenty of brain power has gone into tuning and shaping the simplex algorithm for computer implementation.

A particularly important period in this development followed the announcement of the competing LP algorithm of Karmarkar in 1984. The news surrounding his *interior-point method* drew great attention to linear programming, just at a time when powerful desktop workstations and personal computers were becoming widely available. This combination sent the LP world into hyperdrive. Indeed, Bob Bixby details a million-fold speedup in simplex LP solvers during the period 1987 to 2002. "Three orders of magnitude in machine speed and three orders of magnitude in algorithmic speed add up to six orders of magnitude in solving power. A model that might have taken a year to solve 10 years ago can now solve in less than 30 seconds."[20] Now that is a considerable improvement in LP horsepower. Things have quieted down in subsequent years, but there is every reason to expect further periods of rapid growth with continued research efforts. The National Science Foundation and the Office of Naval Research, in particular, are funding projects to set the stage for a new generation of simplex solvers, aiming to take advantage of the many-core capabilities planned for future computer chips. If the past is any guide, LP researchers will respond to improved hardware with improved implementations, driving Dantzig's simplex algorithm to both faster solutions and larger models.

Behind the Name

The name of the algorithm sounds like a modern corporate term, playing on the word "simple." In fact, the first few hits with a Google search are precisely that. But in geometry a *simplex* is a classical object, namely an n-dimensional polytope having $n + 1$ corner points. That is a plate of undefined terms, but in 2-dimensional space a simplex is just a triangle and in 3-dimensional space a tetrahedron. Dantzig's friend Theodore Motzkin proposed that he borrow this geometric object's name. "The term *simplex method* arose out of a discussion with T. Motzkin who felt that the approach that I was using, when viewed in the geometry of the columns, was best described as a movement from one simplex to a neighboring one."[21] A pity this was adopted. "Dantzig's algorithm" has a nice ring to it and it would be a fitting tribute to his contributions.

Two for the Price of One: LP Duality

An optimal dictionary provides a proof that the simplex algorithm produced a best-possible solution. But we might have a difficult time convincing our manager that the result is in fact correct. Indeed, we probably would not want to ask him or her to go through the entire sequence of pivot steps, checking the arithmetic along the way. Nor could we request an examination of the 100,000 or so lines of computer code in Bob Bixby's LP solver, verifying that it correctly implements Dantzig's algorithm. What we need is a concise route to the dictionary's proof. Providing such a route is the role of *duality* in linear programming.

$$Z = 450 - N - \frac{3}{2}B - \frac{7}{4}S.$$

We obtained this expression from the original definition of Z by adding equations satisfied by all allowable assignments. We know this is true, but unfortunately our manager does not. But not to worry. If in this final Z equation we substitute for N and S their definitions from the original dictionary, then we get back the original expression for total profit. This should convince our manager that we haven't pulled any tricks. There is a cleaner way to do this, however, making a direct argument that earning profit greater than $450 is impossible.

To build the argument, we take the original constraints for nickel and steel and multiply them by the values we find in front of N and S in the final profit equation, not including the negative signs, that is,

$$1 \times (3A + 3B + 1C \le 100)$$

and

$$\frac{7}{4} \times (4A + 2B + 8C \le 200).$$

Carrying out this arithmetic and adding together the resulting two inequalities, we obtain the single inequality

$$10A + 6\frac{1}{2}B + 15C \le 450,$$

which our manager must agree is satisfied by any allowable level of widget production. But comparing this with our total profit $10A + 5B + 15C$, we see that we cannot earn more than $450, even if the profit margin on widget B were increased to $6.50 per unit. A few multiplications and additions allowed us to conclude that our proposed production schedule is optimal.

Let's review the important points in the above argument. First, the values we extract from in front of N and S, call them y_N and y_S, are both nonnegative, allowing us to use them as multipliers for the nickel and steel constraints. Second, when we add the two resulting constraints the values in front of A, B, and C are each at least as large as the profits for the respective widgets.

These rules governing our use of y_N and y_S are in fact the constraints for another LP problem:

$$\text{minimize } 100y_N + 200y_S$$

$$\text{subject to}$$

$$3y_N + 4y_S \geq 10$$

$$3y_N + 2y_S \geq 5$$

$$1y_N + 8y_S \geq 15$$

$$y_N \geq 0, \; y_S \geq 0.$$

The three constraints correspond to the variables A, B, and C, ensuring that the multipliers deliver an inequality with values at least as large as the per unit profits of the three widgets. An assignment of values y_N and y_S satisfying these constraints tells us that profit can be no greater than $100y_N + 200y_S$. Thus, in order to obtain a convincing argument to present to the manager, we seek to minimize this quantity.

The new LP model is called the *dual problem*, and, following a suggestion of Dantzig's father Tobias, the original problem is referred to as the *primal problem*. The constraints of the dual LP are such that any allowable assignment of values to the dual variables provides a bound on the primal objective. In our case we have

$$10A + 5B + 15C \leq 100y_N + 200y_S.$$

The optimal simplex dictionary gives values

$$A = 30, \; B = 0, \; C = 10, \; y_N = 1, \; y_S = \frac{7}{4}$$

that make both sides of this inequality equal to 450, proving that 450 is the optimal value for the primal objective and also that 450 is the optimal value for the dual objective. Two LP solutions for the price of one.

It is remarkable that there always exists such a simple and elegant proof of optimality: the simplex algorithm constructs multipliers that can be used to combine the primal LP constraints into a convincing statement that no solution gives an objective value greater than that supplied by the final dictionary. Moreover, the multipliers are themselves an optimal solution to the dual LP problem and the optimal primal and dual objective values are equal. This beautiful result is known as the *strong duality theorem*, first stated and proved by John von Neumann.[22]

Strong duality gets top billing in LP theory, but in our TSP discussion we really only need the much easier statement that any dual LP solution provides a bound on the primal objective; this is called the *weak duality theorem*. And don't worry if you missed a few details in our rush through material in the past few pages: in the special case of the TSP we provide an intuitive explanation of duality, showing how to trap the salesman with linear inequalities.

The Degree LP Relaxation of the TSP

It didn't take long for LP techniques to enter the TSP world. Dantzig's encounter with Hotelling and von Neumann took place on September 9, 1948, and in the fall of the following year Julia Robinson published the first LP-based method for the TSP.

On the surface, the salesman problem does not appear to fit into the general economic-planning model considered by Dantzig. Indeed, if you put five TSP researchers in a room they will likely come up with five different explanations of why linear programming is natural as a TSP framework. The view I like best is the one mentioned in the opening paragraph of the current chapter, namely, considering LP as a means to obtain quality guarantees by combining simple rules satisfied by all tours. This is straight from the duality playbook: the rules are combined by an assignment of values to dual variables and the guarantee pops out via the weak duality theorem.

Let's take a look at this in action. As usual, we consider the symmetric version of the TSP, where costs do not depend on the direction of travel. The use of graph-theory terminology for such instances should now be familiar, with cities corresponding to vertices and roads corresponding to edges. A tour is a selection of edges that together form a Hamiltonian circuit, as illustrated by the red lines in the 24-city complete graph displayed in figure 5.2.

Grey edges and red edges are fine in a drawing, but to specify a mathematical formulation of the problem we are better off using zeros and ones: the edges assigned value one are those that are in the tour. The 24-city example would have 276 such values. That is way too many to consider without getting trapped in a pile of notation, so let's go back to a more humble six cities, as indicated in figure 5.3. In this case we need fifteen values to specify a tour. Call these values x_{ij} for each pair of vertices

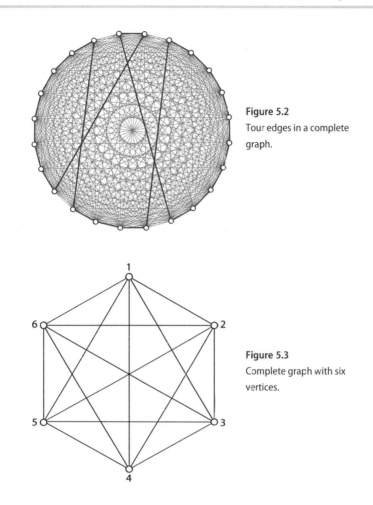

Figure 5.2
Tour edges in a complete graph.

Figure 5.3
Complete graph with six vertices.

i and j, that is, x_{12}, x_{13}, x_{14}, x_{15}, x_{16}, x_{23}, x_{24}, x_{25}, x_{26}, x_{34}, x_{35}, x_{36}, x_{45}, x_{46}, and x_{56}. These are the variables in our LP model, measuring, if you like, the "economic activity" of whether or not the edge joining the pair of vertices is used in a tour.

Denoting by c_{ij} the cost of travel between each pair of cities, the cost of a tour can be written as the linear expression $c_{12}x_{12} + c_{13}x_{13} + \cdots + c_{56}x_{56}$, since c_{ij} multiplied by x_{ij} has the value c_{ij} if the edge is in the tour and the value zero otherwise. This expression is what we minimize in the TSP.

Thus we have our variables and our objective. What about constraints? Oops. We can't say simply that we want the solution to pick out a tour; to use the tools of linear programming we must stick with Dantzig's model and apply linear rules only. Finding such rules is the art and science of our approach.

Degree Constraints

To get things going, the Mad Hatter would note that each of the variables in the LP must be nonnegative. That is fine, but to make real progress we must look into what is special about subsets of edges that form tours. Some subsets are tours, while others are not. How can we make this distinction with linear constraints?

This is the starting point for Julia Robinson's work. She notes that every vertex of the graph meets exactly two edges in any tour. This is not itself a linear rule, but it implies that if we add up the x_{ij} values for all edges meeting a given vertex, then the sum must be exactly two. Thus we have a *degree constraint* for each city, giving us the LP model:

$$\text{minimize } c_{12}x_{12} + c_{13}x_{13} + \cdots + c_{56}x_{56}$$

$$\text{subject to}$$

$$x_{12} + x_{13} + x_{14} + x_{15} + x_{16} = 2$$

$$x_{12} + x_{23} + x_{24} + x_{25} + x_{26} = 2$$

$$x_{13} + x_{23} + x_{34} + x_{35} + x_{36} = 2$$

$$x_{14} + x_{24} + x_{34} + x_{45} + x_{46} = 2$$

$$x_{15} + x_{25} + x_{35} + x_{45} + x_{56} = 2$$

$$x_{16} + x_{26} + x_{36} + x_{46} + x_{56} = 2$$

$$x_{ij} \geq 0 \text{ for each pair of vertices } (i, j).$$

This model is called the *degree LP relaxation* of the TSP.

An optimal solution to the relaxation will itself not typically be a tour, but it nonetheless provides valuable information. Indeed, every tour is an allowable solution to the LP problem, so the optimal LP objective can never be greater than the cost of an optimal tour. This is the most important point to understand. In the LP problem we optimize over a larger set of allowable solutions and we thus obtain a bound on how cheap a tour can possibly be. The bound is a number X such that no tour can have cost less than X, just as we were sure that sales of widgets could not exceed \$450.

Control Zones

This bounding concept is important enough to be worth a second view. So let's take another angle on the degree LP relaxation, this time looking directly at the dual problem. The idea here is a nifty technique for geometric TSP instances introduced by Michael Jünger and William Pulleyblank.[23] In this description we assume our TSP instance consists of a set of points with straight-line distances as travel costs.

To begin, suppose we draw a disk of radius r centered at a city in such a way that the disk does not touch any of the remaining cities, as in figure 5.4. The salesman must at some point in his or her tour visit this city, and to do so he or she will need to travel at least distance r to arrive at the city and at least distance r to leave the city. We can conclude that every tour has length at least $2r$. Moreover, we can draw a separate disk of radius r_i for each city i, as long as the disks don't overlap, as illustrated in figure 5.5. In this way we get twice the sum of the radii of the disks as a bound on the length of any TSP tour. Jünger and Pulleyblank call these disks *control zones*.

We want the control-zone bound to be as large as possible, so we should choose the radii of the disks so as to maximize twice their sum, subject to the condition that disks do not overlap. The nonoverlapping condition can be expressed succinctly as follows: for each pair of cities i and j, the sum of the radii r_i and r_j must be no greater than the distance between the cities, that is, r_i and r_j must satisfy

$$r_i + r_j \leq c_{ij}.$$

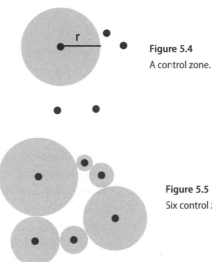

Figure 5.4
A control zone.

Figure 5.5
Six control zones.

Figure 5.6
Left: Michael Jünger. Photograph by Regine Strobl. Right: William Pulleyblank. Photograph by Nick Harvey.

Thus, to obtain the best-possible packing of control zones, we solve the following LP problem:

$$\text{maximize } 2r_1 + 2r_2 + 2r_3 + 2r_4 + 2r_5 + 2r_6$$

$$\text{subject to}$$

$$r_1 + r_2 \leq c_{12}, \ r_1 + r_3 \leq c_{13}, \ r_1 + r_4 \leq c_{14},$$

$$r_1 + r_5 \leq c_{15}, \ r_1 + r_6 \leq c_{16}, \ r_2 + r_3 \leq c_{23},$$

$$r_2 + r_4 \leq c_{24}, \ r_2 + r_5 \leq c_{25}, \ r_2 + r_6 \leq c_{26},$$

$$r_3 + r_4 \leq c_{34}, \ r_3 + r_5 \leq c_{35}, \ r_3 + r_6 \leq c_{36},$$

$$r_4 + r_5 \leq c_{45}, \ r_4 + r_6 \leq c_{46}, \ r_5 + r_6 \leq c_{56}.$$

Any allowable solution to this model gives a nonoverlapping collection of control zones and thus a bound on the length of any TSP tour; an optimal solution to the model gives a strongest-possible control-zone bound.

The above statements should remind you of the weak duality theorem. Indeed, the zone-packing problem is precisely the dual of the degree LP relaxation! To see this, note that the dual problem has a multiplier y_i for each vertex i, corresponding to the ith degree constraint. When we multiply the constraints by the y_i's and add them up, the resulting linear form must have value no greater than c_{ij} in front of each variable x_{ij}. Thus, since the variable x_{ij} appears in the ith constraint and in the jth constraint, we require $y_i + y_j \leq c_{ij}$. Exactly the nonoverlapping condition, with each radius r_i now named y_i.

We must admit that we are cheating a bit, since the dual LP allows control zones with negative radii. These possibly negative values are due to the fact that the degree constraints are equations rather than inequalities, and it is perfectly legal to multiply an equation by a negative number. This is only a technical point, however, since with the triangle inequality one can prove that there exists always an optimal set of dual multipliers such that each value is nonnegative, even though this is not an explicit restriction in the model.

Eliminating Subtours

The next step in the LP approach to the TSP is the introduction of a simple but powerful collection of additional rules. To motivate these rules, consider again the packing of control zones in the six-city example of figure 5.5. This packing creates a ring, making it easy to trace out a tour that moves from zone to zone without any wasted space in between. This unfortunately is not the typical situation. More common is the behavior exhibited in figure 5.7, where zones bump into each other, forcing us to leave a large gap between clusters of cities. It is this gap that is exploited by the new rules.

The geometric version of the idea is the following. In the gap left over from the zone packing we can draw a strip encircling a cluster of points, as illustrated in figure 5.8. Any tour must at some time visit the cluster, so a salesman must cross over the strip at least twice, once on the way in and once on the way out. We can thus add twice the smallest width of the strip to our bound. Jünger and Pulleyblank call such a strip a *moat*, like the waterways surrounding castles.

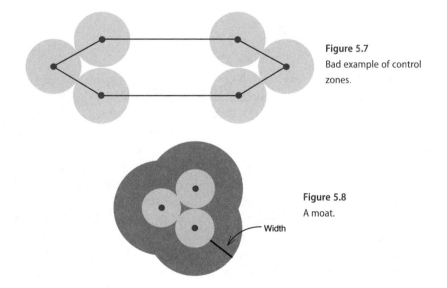

Figure 5.7
Bad example of control zones.

Figure 5.8
A moat.

Width

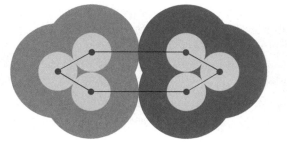

Figure 5.9
Use of moats to fill a gap.

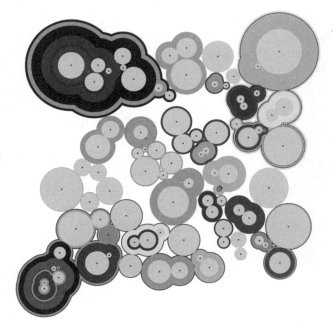

Figure 5.10
Packing of moats and
control zones.

The use of two moats to close the gap in the six-city example is illustrated in figure 5.9. In this case there is again no wasted space as the tour moves from city to city. Although we are not always so fortunate, a careful packing of moats and control zones will often provide a very strong bound. For example, the 100-city packing displayed in figure 5.10 is good enough to prove that a tour is no more than 0.65% longer than optimal.

This 100-city example is pretty, but you can get a better feeling for the bound by playing around with smaller instances. For this I recommend the software package *Geodual* created by Mike Jünger's team at the University of Cologne in Germany.[24] Their package finds optimal tours for TSP instances with up to twenty cities or so, together with beautiful drawings of moats and zones. A sample of their work is displayed in figure 5.11.

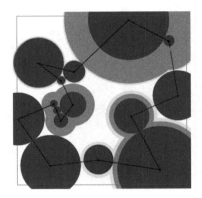

Figure 5.11
Screen shot of Geodual in
action on a 15-city example.

Subtour Inequalities

In a TSP instance such as the six-city example of figure 5.7, the degree
LP relaxation plays a trick on us if we try to use its primal solution
to construct a tour. Indeed, given the large travel costs associated with
the edges spanning the gap, the simplex algorithm will deliver a solution
consisting of two triangles, rather than a single circuit through the six
points. This is an allowable solution to the relaxation, but it is certainly
not allowable for the salesman.

The quick remedy is to note that any tour must include at least two of
the edges crossing the gap between the clusters, that is, at least two of the
green edges indicated in figure 5.12. Analogous to Jünger and Pulleyblank's
moat construction, we can impose the rule that the sum of the variables
corresponding to these inter-cluster edges be at least two:

$$x_{13} + x_{14} + x_{15} + x_{23} + x_{24} + x_{25} + x_{63} + x_{64} + x_{65} \geq 2.$$

Figure 5.12
Edges in a subtour inequality.

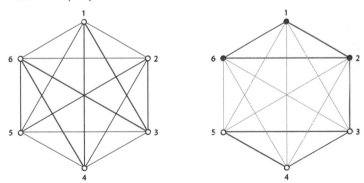

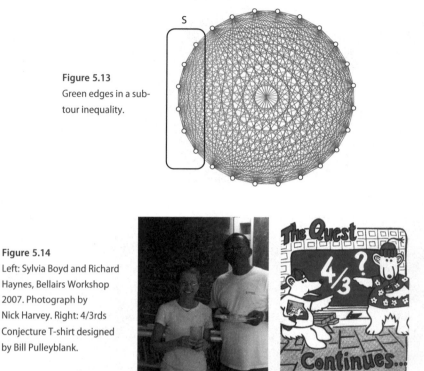

Figure 5.13
Green edges in a sub-
tour inequality.

Figure 5.14
Left: Sylvia Boyd and Richard
Haynes, Bellairs Workshop
2007. Photograph by
Nick Harvey. Right: 4/3rds
Conjecture T-shirt designed
by Bill Pulleyblank.

Combined with the degree constraints, this inequality forbids a solution
that includes the subtours indicated by the red edges in the left-hand side
of figure 5.12. Hence the name *subtour-elimination constraint*, or simply
subtour inequality.

We can create a subtour inequality for any proper subset S of cities,
stating that the sum of the variables corresponding to edges having one
end in S and the other end not in S must be at least two. In figure 5.12 we
have $S = \{1, 2, 6\}$. A larger example is displayed in figure 5.13, where the
set S consists of the vertices in the rectangular region and the variables in
the subtour inequality are those corresponding to the green edges.

These additional rules are simple, but they pack a big punch when combined via linear programming. Indeed, the quality of the bounds obtained
by the *subtour LP relaxation*, consisting of the degree LP relaxation plus all
possible subtour inequalities, is a key ingredient in the success of the entire
LP approach to solving the TSP in practice. For example, the subtour-
relaxation bound is almost always within 1% of the length of an optimal
tour for randomly generated geometric instances. In the specific case of the

42-city USA data set, the bound is 697 units, compared with the optimal tour length of 699 units. In this instance, the tour is only 0.3% longer than the quality guarantee.

The 4/3rds Conjecture

Such strong guarantees do not hold in general, but bad examples seem to be the exception rather than the rule. An interesting question is to determine just how good or bad the bound can be for instances satisfying the triangle inequality. On the positive side, it is known that the cost of an optimal tour is never more than 3/2 times the subtour bound. This is a nice theoretical result, particularly when compared to the theorem of Christofides we cited in the previous chapter.

On the negative side, there is a known family of instances such that the ratio between the cost of an optimal tour and the value of the subtour bound gets closer and closer to 4/3 as the number of cities gets larger and larger. But is this the worst-possible case? The question is known as the "4/3rds Conjecture" and the reigning expert on the topic is Sylvia Boyd from the University of Ottawa in Canada. Together with colleagues, she has verified the conjecture for all instances having at most ten cities and she has posed a sharper question that may provide the angle needed to finally prove the full result.[25]

It might appear to be a small step, but a move from a guarantee of 3/2 to a guarantee of 4/3 will almost certainly require a deep understanding of the structure of the allowable solutions to the subtour LP relaxation.[26] Such an understanding would in turn yield new methods for producing TSP rules that could possibly push the limits of practical computation.

Upper Bounds on Variables

We did not mention this earlier, but the degree LP relaxation is rather special in that there always exist optimal solutions such that every variable has value 0, 1, or 2; we do not need to worry about fractional edges. This is not true for the subtour relaxation, where variables are sometimes assigned nasty-looking fractional values. We have to deal with this in general, but an important special case gets us off to a nice start. Indeed, a variable x_{ij} assigned a value 2 in the degree LP solution corresponds to a subtour that goes from city i to city j and then immediately back to city i. The subtour inequality determined by $S = \{i, j\}$ will forbid such an assignment, but it is slightly more efficient to deal with this via the simple rule $x_{ij} \leq 1$. Imposing such upper bounds on all variables gives a better starting point than the degree LP relaxation alone, and this is what is typically used in practice. Moreover, for this improved starting model there exists always an optimal solution such that every variable is either 0, 1/2, or 1. Not quite integer valued, but pretty close.

A Perfect Relaxation

Subtour inequalities bound a salesman solution to within a small factor of optimality, but can we hope for much better results with additional rules? Good grief, yes! It takes a bit of mathematics to get there, but you will soon see what I mean.

The Geometry of Linear Programming

Up to this point we have been treating linear programming as a purely algebraic topic, dealing with variables, equations, and their manipulation. This is unfair to George Dantzig and most LP researchers, who view their subject as one possessing its own geometric elegance.

To see what we have been missing, let's consider a small example:

$$\text{maximize } x + 2y$$

$$\text{subject to}$$

$$x + y \leq 13$$

$$x \leq 8, \ y \leq 8$$

$$x \geq 0, \ y \geq 0.$$

An allowable solution to this model can be considered as a point (x, y) in two-dimensional space, where the horizontal axis indicates the value of x and the vertical axis indicates the value of y.

The full set of allowable points is called the *feasible region* of the model. To obtain a view of this region, focus first on the single constraint $x + y \leq 13$. The corresponding line $x + y = 13$ separates points (x, y) into two sets, those on the forbidden side of the line and those on the allowable side, where points on the line itself are also allowable. The allowable side is called a *halfspace* and it is indicated in two ways in figure 5.15; on the left by red arrows pointing in the allowable direction and on the right by red shading.

The feasible region for the example LP problem consists of the intersection of five halfspaces corresponding to the problem's five constraints,

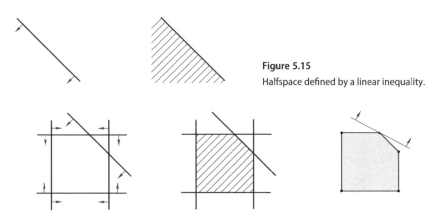

Figure 5.15
Halfspace defined by a linear inequality.

Figure 5.16
Geometric view of an LP problem.

as displayed in figure 5.16. In this geometry, the LP problem is to push the line of the objective, indicated by the blue line in the right-hand side of the figure, as far up as possible, while still hitting the feasible region. Thus the red point, $(5, 8)$, is the optimal solution in this example.

Note that the optimal point is one of the five corners of the feasible region; the others are $(8, 5)$, $(8, 0)$, $(0, 0)$, and $(0, 8)$, going clockwise around the border. An important observation is that no matter what objective is set for the LP model, we can be sure that one of these five corner points will be an optimal solution. Indeed, changing the objective means a change in the slope of the blue line, but if we push any line as far as we can it will always meet one of the corners just before leaving the feasible region.

This is a very nice general property. It means that an LP problem, which has an infinite collection of allowable solutions, can be solved by considering only the finite number of corner points. Indeed, the simplex algorithm can be viewed as a method for moving from corner to corner along the edges of the feasible region.

You might ask why we bother with the simplex algorithm when an LP problem can be solved by creating a list of corners. It comes down to numbers, keeping in mind the quote from George Dantzig: "For a big problem there could be many combinations (corner points)—perhaps as many as the stars in the heavens." A corner is determined by intersecting the bounding planes for a collection of d halfspaces when we are in d dimensions. Not every such intersection will lie in the LP feasible region, but enough of them will be allowable to cause a headache for anyone trying to produce a full list.

Minkowski's Theorem

This discussion of corner points for LP problems is fine, but for the TSP the geometry is the other way around, that is, we have already a full list of points, corresponding to the tours in the graph. We have the list, but we do not have an LP feasible region. In this view the search for TSP rules is a search for halfspaces that trap in the tours.

Let's look at this more closely. Every tour for a six-city TSP corresponds to a point in 15-dimensional space, that is, one dimension for each pair of cities. Each tour point consists of zeros and ones, with the ones indicating the tour edges. This large set of zero/one-valued points is sitting in 15-dimensional space, but without any geometric structure to allow us to pick out the point corresponding to a shortest-possible tour. It is the role of linear programming to provide this missing structure.

We cannot draw pictures of 15-dimensional space, so consider instead a similar problem in two dimensions, that is, from a list of (x, y) points we would like to choose one having the greatest value for some specified objective. In this case we are looking for linear inequalities satisfied by every point on the list, or, in other words, halfspaces with all of our points on the allowable side. The process of building inequalities tailored for such a problem is illustrated in figure 5.17; the six halfspaces encircle our point set, and each corner of the enclosed region is itself one of our points. This

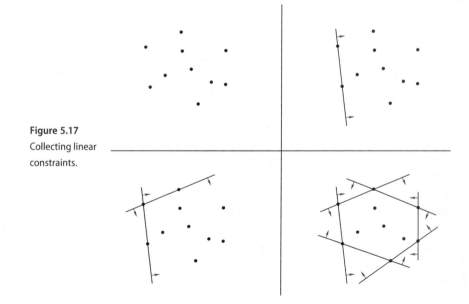

Figure 5.17
Collecting linear constraints.

Figure 5.18
The convex hull.

is great news from the LP perspective: the simplex algorithm applied to the region will deliver an optimal solution to our problem.

The construction of the perfectly fitting region, drawn again in figure 5.18, is not just a lucky case. It is in fact possible to enclose any set of points in an LP feasible region such that each corner of the region is a point in the set. In two dimensions you can think of stretching out a rubber band and letting it snap back to enclose the points. In three dimensions imagine shrink-wrapping the points, again giving a perfect fit. Going on to higher dimensions is tough to visualize, but the result, described by Hermann Minkowski at the turn of the twentieth century, is not difficult to prove by making a connection between algebra and geometry.

The perfectly fitting enclosed region for a set of points is called the *convex hull* of the set. In two-dimensional space a convex hull is a special type of solid polygon and in three dimensions a convex hull is a faceted object like the Platonic solids or a cut diamond. The objects in general are called *convex polytopes* and they have long been the subject of study by mathematicians. The text *Lectures on Polytopes* by Günter Ziegler gives a fantastic technical presentation of this research area; the details in the book are advanced, but the introduction and first few chapters will give you a good idea about what keeps mathematicians interested in these classical geometric structures.[27]

The TSP Polytope

The upshot of Minkowski's theorem is no less than the fact that any TSP can be modeled precisely as a linear-programming problem! This puts the search for TSP rules on firm theoretical ground: the necessary linear inequalities are out there, we just have to find them.

Now, before you get too excited, I must point out a potential difficulty in that the number of halfspaces one needs to describe a TSP polytope is enormous. By the time we get up to ten cities it is known that at least 51,043,900,866 inequalities are required.[28] Still, numbers alone will not defeat us. Harold Kuhn emphasizes this point in notes from a George Dantzig Memorial Lecture delivered in 2008; in the text Kuhn refers to his own TSP studies carried out in 1953.

I had a number of contacts with George throughout the summer discussing this and other problems. And I know that George attended my lecture at the end of the summer (as did Selmer Johnson, Ray Fulkerson, and Alan Hoffman). We were both keenly aware of the fact that, although the complete set of faces (or constraints) in the linear programming formulation of the Traveling Salesman Problem was enormous, if you could find an optimal solution to a relaxed problem with a subset of the faces that is a tour, then you had solved the underlying Traveling Salesman Problem.

What we need is a good understanding of how to work with the inequalities, that is, how to produce an inequality when it is useful for the TSP instance we are trying to solve. This gets to the heart of the LP approach to the TSP, as we discuss in chapter 6.

Integer Programming

If the linear-programming users of the world could be granted a wish, they would with one voice shout out for whole-number solutions from the LP genie. An obvious reason is to avoid nastiness like dealing with a production schedule calling for $33\frac{1}{3}$ widgets. But the main point is to capture decisions that come down to individual choices. Should we build an additional factory? Yes or no. Should we bring a new product to market? Yes or no. Such decisions can be brought into an LP model if we are permitted to include variables that take on values zero or one only, no fractions accepted. This is a powerful extension of linear programming, but one that for the present comes at great computational cost.

The restriction to integers does not fit into Dantzig's theory and it cannot be directly handled by the simplex algorithm or other LP methods. Thousands of LP users nonetheless go ahead and include such restrictions in their models everyday, unable to resist the flexibility that integer-only variables bring to the table. The extended framework is known as *integer programming*, or *IP* for short.

Dantzig himself was the first to document how versatile integer programming can be. In a paper that is fundamental in both the field of optimization and the field of complexity theory, he showed how each member of a long list of important optimization problems can be modeled as an IP problem.[29] Dantzig described his work as follows in his 1963 LP book.[30]

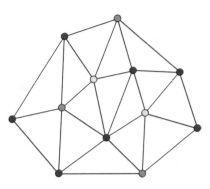

Figure 5.19
Four-coloring of a graph.

Our purpose is systematically to review and classify problems that can be reduced to linear programs, some or all of whose variables are integer valued. We shall show that a host of difficult, indeed seemingly impossible, problems of a nonlinear, nonconvex, and combinatorial character are now open for direct attack.

His problem classes include the TSP and optimally coloring maps.

The coloring problem is a good example to see integer programming in action. Rather than coloring a map, let's consider the more general problem of assigning colors to the vertices of a graph in such a way that any two vertices joined by an edge receive different colors. This captures the map problem by placing a single vertex in each region of the map and creating an edge joining two vertices if their regions share a border.

An example of a graph with a four-coloring is displayed in figure 5.19. It is easy to see that this particular graph cannot be colored with three colors, but in general it is a difficult problem to determine if three colors suffice. Let's set it up as an IP model. For each vertex i we create three nonnegative variables, $x_{i,red}$, $x_{i,green}$, and $x_{i,blue}$, with the interpretation that $x_{i,red}$ is set to 1 if vertex i receives the color red and $x_{i,red}$ is 0 otherwise, and similarly for green and blue. Since we must assign vertex i one of the three colors, we have the constraint

$$x_{i,red} + x_{i,green} + x_{i,blue} = 1.$$

Now for any edge (i, j) the vertices i and j cannot both be assigned the same color, so we have the three constraints

$$x_{i,red} + x_{j,red} \leq 1$$

$$x_{i,green} + x_{j,green} \leq 1$$

$$x_{i,blue} + x_{j,blue} \leq 1.$$

An integer-valued solution to this system of equations and inequalities gives a valid coloring with red, green, and blue. Note, however, that an LP solution to the model without the integer restrictions would be to set all variables equal to 1/3, providing no information that could be used to assign colors to the vertices. The integer variables rule the day.

The TSP as an IP

In one sense we have already modeled the TSP as an integer-programming problem. Indeed, with integer variables, the combination of the degree constraints and the subtour inequalities is enough to ensure that any solution will be a tour. This is an IP model, but it is not one that could be handed over directly to an IP solver. The difficulty is that the collection of subtour inequalities numbers approximately $2^{n/2}$ for an n-city TSP.

For this reason Dantzig described an alternative model of the TSP, using n^3 nonnegative variables but only $n + n^2$ constraints. The idea is that rather than specifying only whether or not we use an edge between two cities, we also specify its position in the tour. Other early TSP researchers, such as Albert Tucker, discovered IP models with even fewer variables and constraints, but we do not want to emphasize these as practical tools for the salesman. Thus far, all alternative models are dominated, in terms of practical performance, by the subtour formulation and the methods we describe in the next chapter. The existence of compact models should, however, convince you that solving IP problems is in general difficult: a polynomial-time IP solver would give directly a polynomial-time TSP solver.

IP Solvers

The difficulty of solving general IP models has not stopped the world from creating them over and over again in numerous business applications. Like in the case of the TSP, it is of no use to throw our hands in the air and tell the world that general IP may be impossible to solve. The problem needs to be addressed in an algorithm-engineering approach. And indeed it is. This work is what often pays the bills for TSP researchers: nearly all known general IP methods were discovered first in pursuit of solutions for the salesman.

For the business IP models created worldwide, several very sophisticated solvers duel in the commercial marketplace. These computer codes have seen great advances in practical performance over the past two decades, and there should be more improvements to come, as we better understand how to tackle the salesman and its IP brother.

Operations Research

Linear and integer programmers can be found in departments of mathematics, computing, business, science, and engineering. The discipline most closely associated with LP and IP, however, is one called *operations research*.

Operations research traces its roots to work on military planning in the mid-1900s, hence the "operations" part of the name. There are now departments and centers of study throughout the world. In the United States, operations research programs can be found at Berkeley, Carnegie Mellon, Columbia, Cornell, Florida, Georgia Tech, Lehigh, Michigan, MIT, Northwestern, Princeton, Rutgers, Stanford, and many other universities.

The name operations research is perhaps not one a marketing agency would suggest. Actually, a slogan for the field developed by a marketer is "The Science of Better." This was the centerpiece of a promotional campaign run by the professional society INFORMS (Institute for Operations Research and Management Science). The campaign material answers the question "What is operations research?" as follows: "In a nutshell, operations research (O.R.) is the discipline of applying advanced analytical methods to help make better decisions." This sums up the field very nicely, bringing home the point that the use of operations-research techniques cuts across industries, from health care to transportation, from finance to forestry. Any place where decisions are made, operations research can be applied. In operations research studies, optimization tools, such as linear and integer programming, are combined with modeling techniques drawn from probability theory, game theory, and elsewhere.

Figure 5.20
Michael Trick, 2010.

To get a feeling for the excitement and breadth of the field, there is no better place to start than the writings of Carnegie Mellon Professor Michael Trick. Trick, a former president of INFORMS, maintains a lively blog on everything OR, including his specialty, operations research applied to sports scheduling.[31] Trick has a great talent for spotting practical opportunities for optimization techniques, so keep an eye on his blog for new applications of the TSP.[32]

6: Cutting Planes

Starting from a solution worked out with strings on a model (which was in fact optimal), Dantzig, Fulkerson, and Johnson had nevertheless to face the possibility that billions of cuts might be needed.
—Alan Hoffman and Philip Wolfe, 1985.[1]

The linear-programming relaxations associated with the traveling salesman problem are wildly complex: the simplex method is no match for problems with constraints numbering in the billions. Fortunately, Dantzig, Fulkerson, and Johnson put forth an elegant idea for handling such complexity. Their cutting-plane method does not attempt to solve the full LP problem in a stroke, but rather computes LP bounds on a pay-as-you-go basis, generating specific TSP inequalities only as they are needed. This is a game changer, and not for the salesman alone.

The Cutting-Plane Method

The road to Dantzig et al.'s tour through the United States begins with the degree LP relaxation and its solution displayed in figure 6.1. The edges drawn in red carry the LP value 1/2 and the edges drawn in black carry the value one, with all other variables assigned the value zero.

You see directly that the simplex algorithm does not lie: every city meets edges of total value two, that is, either two black edges or two reds and one black. You also see that the solution is certainly not a tour. One obvious point is the island of four cities in the northeast, where the split salesman refuses to take any of the roads leading out of the island. This difficulty will disappear if we roll out the full set of subtour-elimination constraints, but that means adding 2,199,023,254,648 inequalities to the model. A tall order for an LP solver.

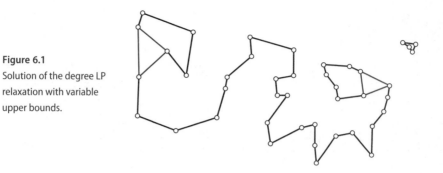

Figure 6.1
Solution of the degree LP relaxation with variable upper bounds.

There is no denying the RAND team's talent for by-hand calculations, but they obviously did not handle directly two trillion subtour inequalities. No, their approach was much more subtle: using the LP solution as a guide, they set out to find just enough inequalities, valid for all tours, so that the simplex algorithm returns an optimal solution that is itself a tour. Since all tours are potential solutions to the LP model, the one identified by the simplex algorithm must be optimal.

To get things rolling, as a first step we select the single subtour-elimination constraint corresponding to the northeastern island, as illustrated in the first drawing in figure 6.2. After adding the inequality to the LP model, the simplex method pops out the solution indicated in the second drawing. Another island has appeared, so we stamp it out with a second subtour inequality, corresponding to the seven cities in the Great Lakes region. The simplex method responds this time with a new solution that contains, yet again, an island of four cities in the center of the country. This may seem like plugging holes in an aging dam: whenever we repair one leak another appears. In fact, the original draft of the Dantzig et al. paper stated that the process was called the "finger in the dike method" by fellow RAND researcher Edwin Paxson.[2]

But looks are deceiving. Although the solution may appear to be no closer to a tour, we are nonetheless making great progress. The initial LP model delivered a solution with objective value 641, but adding the single northeastern subtour inequality raised this to 676 and the second inequality further raised the bound to 681. We are clearly on the right track, despite the pesky islands appearing in each new solution. The next five steps, indicated in figure 6.2, continue with bounds of 682.5, 686, 686, 688, and 697, respectively.

The final solution in the sequence has value only two units less than the length of an optimal tour. This is great, but how to continue? There are no islands in the solution, but that in itself is not a problem. Indeed, if you

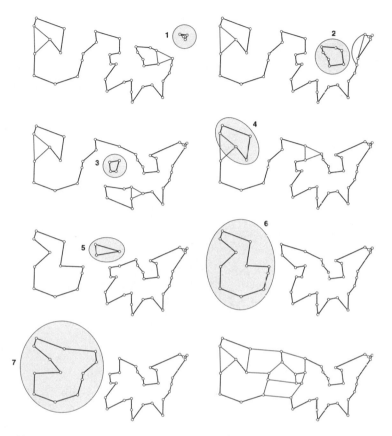

Figure 6.2
First seven steps of the cutting-plane method.

look carefully at figure 6.2 you see that in step four we also had a connected solution. In that case there was a cluster of cities having a total value of one going from the cluster to the remaining cities, that is, exactly two red edges. No such set is apparent in the final LP solution. In fact, we will see later that this solution actually satisfies every subtour-elimination constraint. This by itself is quite interesting. It means that we have computed an optimal solution to the full set of over two trillion subtour-elimination constraints by adding only seven inequalities! Quite interesting, but not sufficient to solve the 42-city instance of the TSP. We somehow have to raise the bound to 699, the length of the optimal tour.

Okay, we have run out of useful subtour-elimination constraints, but there are many, many other inequalities to choose from in a description of the 42-city TSP polytope. We just need to find one that is not satisfied by our LP solution. Dantzig et al. used creative ad hoc arguments to formulate

Figure 6.3
Configuration of four sets
that must be crossed at
least ten times.

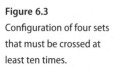
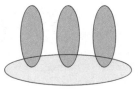
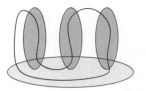

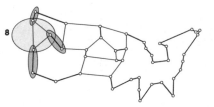

Figure 6.4
Eighth and ninth steps of the
cutting-plane method.

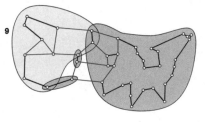

two such constraints, but we can replace these with a general rule involving four subsets of cities, arranged as in the Venn diagram indicated in the left-hand side of figure 6.3. Playing around with possible tour routes such as the one indicated in the right-hand side of the figure, you should be able to convince yourself that every tour must cross the four boundaries of the sets at least ten times. In other words, every tour must satisfy the constraint obtained by adding together the four subtour inequalities and replacing the right-hand side by the value ten.[3]

Equipped with this four-set configuration, we can finish off the USA problem. As our eighth constraint we take the inequality corresponding to the sets indicated in the first drawing in figure 6.4. The border of the yellow set is crossed by three black edges, the borders of two of the blue sets are each crossed by one black edge and two red edges, and the border of the third blue set is crossed by four red edges. Summing these up, the borders of the four sets are crossed by five black edges and eight red edges, for a total LP value of nine.[4] This is less than the required ten crossings, so we add the four-set inequality to our LP model, resulting in the solution of value 698 displayed in the second drawing in figure 6.4. Adding a second four-set configuration wraps thing up: the simplex algorithm delivers a tour. At this point we bring in the dual LP solution to convince one and all that 699 is indeed the optimal value for this instance of the TSP.

Nine inequalities together settle the challenge of routing a salesman through the United States. That is amazing. Nine inequalities from trillions and trillions. The RAND team had tremendous intuition to even attempt such a step-by-step process, particularly when you realize the amount of work needed to carry out computations with by-hand calculations.

The chosen nine inequalities are called *cutting planes*, since at each step the corresponding halfspace cuts off the current LP solution from the feasible region of the model. The full procedure is called the *cutting-plane method*; less flowery than Paxson's suggestion, but also less pessimistic.

Dantzig et al. conclude their famous paper with the following modest remark.[5]

> It is clear that we have left unanswered practically any question one might pose of a theoretical nature concerning the traveling-salesman problem; however, we hope that the feasibility of attacking problems involving a moderate number of points has been success-fully demonstrated, and that perhaps some of the ideas can be used in problems of similar nature.

Successful indeed! Repercussions of the work are still being felt in the world of applied mathematics.

A Catalog of TSP Inequalities

The first of the two non-subtour inequalities employed by Dantzig et al. is a disguised form of the first of our four-set configurations. Their second non-subtour inequality was quite different, however, and a footnote points to Irving Glicksberg as an unsung hero in the effort: "We are indebted to I. Glicksberg of Rand for pointing out relations of this kind to us."[6]

Knowing it may be best not to rely on ad hoc arguments, even those produced by their friend Glicksberg, Fulkerson composed a letter to TSP expert Isidor Heller, dated March 11, 1954; the object he refers to as "C_n" is the convex hull of tours for an n-city problem.

Figure 6.5
George Dantzig, Ray Fulkerson, and Selmer Johnson. Courtesy of the National
Academy of Engineering, Mrs. Merle Fulkerson Guthrie, and the University of
Texas Center for American History, respectively.

Recently, G. Dantzig, S. Johnson, and I have been working on
computational aspects of the problem via linear programming techniques even though we don't know, of course, all the faces of the
convex C_n of tours for general n. The methods we have been using
seem hopeful, however; in particular, an optimal tour has been
found by hand computation for a large scale problem using 48 cities,
rather quickly. We have found it convenient in translating Dantzig's
simplex algorithm in terms of the map of points, to identify tours
which differ only in direction of traversal. For example, C_5 can
be characterized by a system of 25 hyperplanes in 10 dimensional
space. We don't know very much about C_n in general, but thought
we might learn more from reading your papers, if they are available.

Similar requests were sent from Dantzig to Harold Kuhn, March 11, 1954,
and from Dantzig to Albert Tucker, March 25, 1954. It is clear the RAND
researchers were actively seeking information on the structure of the TSP
polytope, to better equip their cutting-plane method.

Comb Inequalities

Dantzig et al. called for help, but the research community was slow in
responding. Being far ahead of your time does have its disadvantages.
Vašek Chvátal finally picked up the theme in the early 1970s; his work on
comb inequalities started the research ball rolling again, nearly two decades
after the RAND study.[7] Martin Grötschel and Manfred Padberg followed
quickly with an extension and analysis of combs that served as a template
for future studies.[8]

By a *comb* we refer to a collection of subsets of vertices arranged as in the Venn diagram displayed in figure 6.7. The yellow set, called the *handle* of the comb, intersects each of the disjoint blue sets, called the *teeth* of the comb. We require that the number k of teeth be odd and at least three. It takes some care to cover all possible cases, but one can show that every tour must cross the borders of the comb at least $3k + 1$ times, extending the observation on four-set configurations.

To get a feeling for the $3k + 1$ requirement, take a close look at the drawings in figure 6.8. The top image shows how a tour could possibly trace through the vertices in a five-tooth comb. Counting the border crossings you see they total 16, that is $3 \times 5 + 1$. In the bottom figure we trace through a configuration having six teeth. This is a no-no, since six is not an odd number, and the tour crosses the borders only 18 times, one less than the $3k + 1$ rule would require. In this even case we were able to pair up the teeth, avoiding the extra exit from the handle.

We have seen that combs finish off the 42-city USA data set. In a second demonstration, Grötschel set a new TSP record with an optimal tour through 120 cities in Germany. Grötschel's hand drawing of an LP solution with possible cutting planes is displayed in figure 6.9. It is great to see this original work: 1/2-valued edges are indicated by wiggly lines, violated subtour inequalities are indicated by red enclosures, and violated

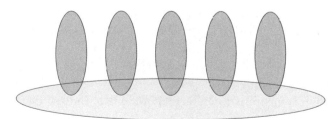

Figure 6.7
Venn diagram of a
comb with five teeth.

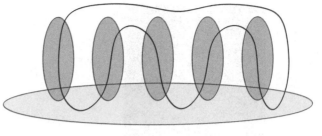

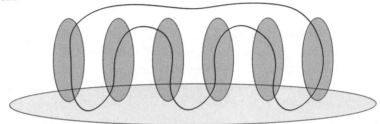

Figure 6.8
Tour crossing
borders of five teeth
and six teeth.

combs are indicated by blue enclosures for the handle and teeth. In each
round of his computation Grötschel added multiple cutting planes, which is
highly recommended when facing large instances of the salesman problem.

Facets of the TSP Polytope

Grötschel's computational success served as a call to the research community to push ahead with a study of further classes of TSP inequalities. The
first results in this effort were obtained by Grötschel himself together with
William Pulleyblank, extending combs to multiple-handle configurations.[9]
Their structure is called a *clique tree* due to the tree-like nature of the
permitted Venn diagrams, such as the example displayed in figure 6.10.

The work of Grötschel and Pulleyblank was followed by other groups
with ever more wild-looking constructions. The inequalities may be wild,
but the work was guided by the fundamental structure of the TSP polytope.
This particular geometry is tough to describe in the high-dimensional
world of the TSP, but the idea is clear when we step back to two
dimensions.

Consider the convex hull displayed in figure 6.11. Each of the indicated
halfspaces touch at least one point in the set, but everyone would agree that
the halfspace at the top is the more fundamental of the two. Indeed, there

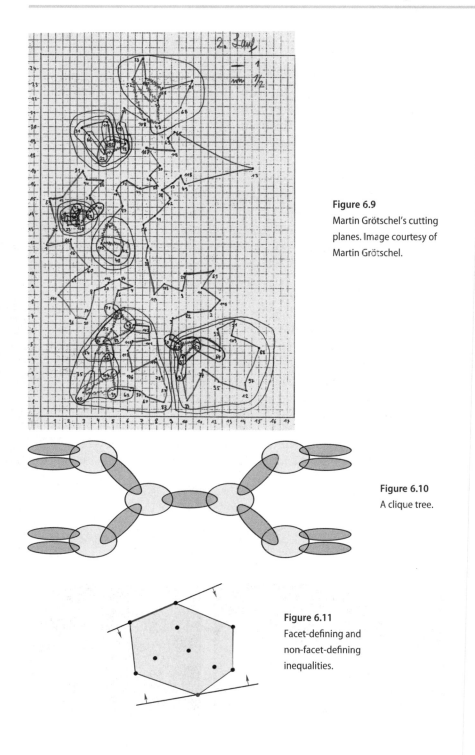

Figure 6.9
Martin Grötschel's cutting planes. Image courtesy of Martin Grötschel.

Figure 6.10
A clique tree.

Figure 6.11
Facet-defining and non-facet-defining inequalities.

are only six halfspaces of this type and they provide a full description of the convex hull. The six halfspaces are called *facet-defining* inequalities, and the six borders they create are called the *facets* of the polytope.

We cannot draw a nice picture of the TSP polytope, but the notion of a facet carries over to the high-dimensional setting.[10] The facet-defining TSP inequalities, together with the degree constraints, provide a complete description of the convex hull of tours. And each of these inequalities must be included in any such description. Thus, although we do not know the full structure of the polytope, we can say for certain that particular inequalities must be part of any complete list that perfectly traps in the set of tours.

Harold Kuhn and others studied facets of small TSP polytopes in the 1950s, but it was Grötschel and Padberg who first proposed that the community focus on facet-defining inequalities when building a catalog of potential cutting planes.[11]

> Our interest in establishing this fact is twofold: Firstly, it is of mathematical interest to know which ones of the proposed inequalities *really matter* in defining this incredibly complex polytope. Secondly, facets are "strongest cutting planes" in an integer programming sense and it is thus natural to expect that such inequalities are of substantial computational value in the numerical solution of this hard combinatorial optimization problem.

They showed subtour inequalities and comb inequalities are facet defining, and Grötschel and Pulleyblank proved that clique-tree inequalities too are in this elite club.

Figure 6.12
Left: Egon Balas, Suzy Mouchet-Padberg, Harold Kuhn, Manfred Padberg, and Martin Grötschel, Berlin, 2001. Courtesy of Martin Grötschel. Right: Giovanni Rinaldi and Denis Naddef, Aussois, 2008. Courtesy of Uwe Zimmermann.

The hunt for facets was pursued through the 1990s, with Denis Naddef of Grenoble and Giovanni Rinaldi of Rome leading the charge.[12] This work has provided a wealth of information that has not yet been fully exploited in computational studies, but our knowledge of the TSP polytope is far from complete: the general theory explains only a small fraction of the fifty-one billion facets that are known for the ten-city polytope. Putting a positive spin on this, we know that much remains to be discovered. A target-rich environment for future TSP work.

The Separation Problem

A large catalog of inequalities is at the disposal of anyone ready to take on the TSP, but using the catalog effectively is not an easy task. Indeed, this is a point where more attention is needed if we are to push the salesman to new heights.

The task is to find from among the known TSP inequalities one or more that are violated by a specified LP solution. This is called the *separation problem*, since we think of the corresponding halfspaces as "separating" the solution from the convex hull of tours. Separation is the heart of the cutting-plane method: computer codes such as Concorde are basically wrappers for calling separation routines. If you want to join in on the TSP party, there is no better topic to study than fast and efficient separation methods.

Maximum Flow and Minimum Cuts

The worker bees of the cutting-plane method are the subtour-elimination constraints, and in this case, at least, the separation problem is well understood. The techniques employed here date back to Cold War–era mathematicians, who studied the movement of equipment through the Eastern European rail network on one side and efficient bombing campaigns to destroy the network on the other side.[13]

Rather than moving to Europe, let's stick with our 42-city TSP, examining again the LP solution obtained after the addition of the first seven cutting planes. In the corresponding LP graph displayed in figure 6.13, we have indicated four paths from Phoenix in the south to Montpelier in the north. Think of sending some item, such as oil, along each of these paths, where through each edge we are permitted to send at most its x_{ij} value, that is, x_{ij} measures the capacity of a pipe between i and j. Sending a value of

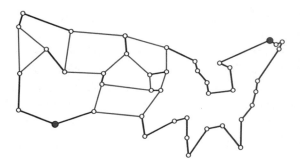

Figure 6.13
Four paths between Phoenix
and Montpelier.

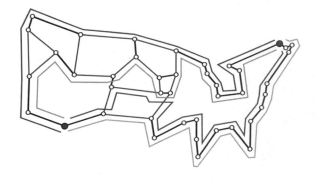

1/2 along each of the four paths, we have a *flow* of total value two between the source and destination.

The dual concept to a flow is a *cut* separating two vertices, that is, a set of edges whose deletion will split the graph into two islands, one containing the source and the other containing the destination. Summing the capacities of the edges in a cut gives a bound on the maximum possible flow, since all oil must at some point move from one island to the other. In the 42-city example we can take as a cut the set of edges meeting Phoenix; the capacity of this cut is two, which, not coincidentally, matches our flow value.

The main theorem of flows and cuts is a "strong duality" result, that is, for any graph and any choice of source and destination, the maximum value of a flow is equal to the minimum capacity of a cut. Furthermore, standard polynomial-time algorithms to compute the maximum flow also compute a minimum cut.

Note that the capacity of a cut is precisely the evaluation of the subtour inequality corresponding to one of the two islands that are created. Thus, the flow of value two between Phoenix and Montpelier implies there are no violated subtour inequalities corresponding to sets containing one of these cities but not the other. Repeating this construction for each of the remaining forty choices for the destination vertex demonstrates that there are no violated subtour inequalities at all.[14]

In general, we solve the subtour-separation problem by choosing a source vertex and computing a maximum flow to each other vertex, that is, by solving $n - 1$ maximum-flow problems. If each of the results is of value two, then we can be sure there are no violated subtour inequalities. On the other hand, if any of the results is less than two, then the corresponding minimum cut yields a violated inequality.[15] It might sound time consuming, but the process can be made to run very fast.

Comb Separation

Subtour-elimination constraints? No problem. Comb inequalities? Not so good. At the moment there is no known polynomial-time algorithm that is guaranteed to spot a violated comb inequality if one exists. The comb-separation problem is also not known to be in the $\mathcal{NP}$-complete complexity class, so its status is wide open. A great need coupled with inconclusive results translates into an important research problem.

This is not to say that combs are currently skipped over in computations. By hook or by crook computer codes must generate combs to push LP bounds after subtour-elimination constraints fail. This is accomplished by hit-or-miss strategies for growing handles, shrinking teeth, splicing sets, and a host of other operations. This can get messy, but at the start violated combs are actually quite easy to spot. Indeed, the form of the inequality suggests that as potential teeth we should consider sets having borders that are crossed by a value close to two in the current LP solution. A ready supply of such potential teeth are sets $S = \{i, j\}$ where $x_{ij} = 1$ in the LP solution, that is, the ends of black edges in the LP graph. So, as a first step we can delete all black edges and examine the remaining red islands. If the borders of any of these are met by an odd number of deleted black edges, then we have a violated comb inequality: the island is the handle and the black edges are the teeth. The process is illustrated in figure 6.14, where we find two potential comb inequalities in the solution of the subtour LP relaxation; in our computation we chose the one having three teeth.

Besides the quick method we just described, there are a number of fancier heuristic algorithms for combs with single-edge teeth, as well as an exact-separation method that runs in polynomial time. These techniques are quite successful and, in true algorithm-engineering fashion, researchers have exploited their success in hit-or-miss algorithms for general combs. The idea is to replace clusters of cities by single vertices and then search for single-edge combs in the shrunken graph. Any combs that are found

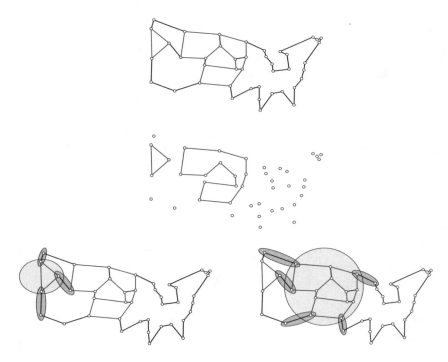

Figure 6.14
Two combs from red islands.

can be expanded back to general combs in the original. The art here is to select clusters in a creative way, typically using methods that search for sets having borders of value close to two in the LP solution. It was a shrinking heuristic of this type that delivered the large three-tooth comb that completed our 42-city computation.

Non-crossing LP Solutions

The search for comb-separation algorithms might appear to be a dirty business, but it can lead to very interesting mathematics, bringing in methods and structures from various fields of study. A nice example is the work of Lisa Fleischer and Éva Tardos at Cornell University in the late 1990s.[16] In an area dominated by the immediate algorithm-engineering needs arising in computational studies, the two researchers took a step back and achieved an elegant and precise theoretical result, rather than a hit-or-miss heuristic algorithm, the first of its kind for general combs.

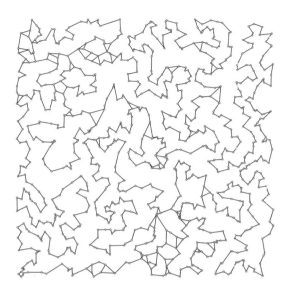

Figure 6.15
Solution of subtour LP relaxation for a 1,000-city instance.

The main idea in the Fleischer-Tardos work is to focus attention on LP solutions that can be drawn as graphs without crossing edges, such as the 42-city solutions we saw earlier or the 1,000-city solution displayed in figure 6.15. This is a nice restriction. Geometric instances of the TSP often have LP solutions that are non-crossing, or nearly so, and this structure can be exploited by tools from the theory of graphs that are not available in the general case. The Fleischer-Tardos analysis yields a separation algorithm that runs in polynomial time, but we must be careful in stating a qualification on the combs that are delivered.

Every tour crosses the boundaries of a k-toothed comb at least $3k + 1$ times. It is also true that any LP solution satisfying all subtour-elimination constraints will evaluate to at least $3k$, that is, a comb inequality can be

Figure 6.16
Lisa Fleischer. Éva Tardos. Adam Letchford.

violated by at most one. Now, presented with a non-crossing LP solution such that there exist combs that evaluate to $3k$, the Fleischer-Tardos algorithm is guaranteed to produce one. The method may, however, fail to spot combs violated by $1 - \delta$ for a small number δ. This is a mixed blessing. On the one hand it is good news that the algorithm will find a comb having maximum violation, but on the other hand it is bad news, from a practical view, to miss out on the many other combs that may exist when the algorithm fails.

It is an open research question to construct a polynomial-time algorithm to solve the full comb-separation problem for non-crossing LP solutions. Such a result would be an important theoretical achievement and it would likely have a direct practical impact on TSP computation. Exploring this question, Adam Letchford of Lancaster University in England took the Fleischer-Tardos ideas in another direction, developing a new class of constraints designed from the ground up to permit polynomial-time separation in the non-crossing case.[17] Letchford's constraints are called *domino-parity inequalities* and they include combs and many more structures as well.

This "many more" sounds like a good thing, but the practical consequences were not so clear. From numerous computational studies we know that comb inequalities pack a big punch, but Letchford's polynomial-time algorithm delivers more general constraints, sometimes not facet defining, even if violated combs are available. The Canadian team of Sylvia Boyd, Sally Cockburn, and Danielle Vella cleared up this matter several years later with an impressive study, combining a computer implementation with by-hand computations to show that Letchford's algorithm works like a charm on modest-sized test instances.[18] Their work was followed by a large computational study led by Daniel Espinoza and Marcos Goycoolea at Georgia Tech, fully automating the algorithm.[19] The end result was an important new module for the Concorde code that played a big role in the sprint up to the record 85,900-city solution.

Edmonds's Glimpse of Heaven

Better inequalities lead to better bounds, and better bounds lead to faster computer codes. But can this route lead to the million-dollar prize? There is a precedent in the stunning solution of the perfect-matching problem by Jack Edmonds.[20]

A perfect matching in a graph is a set of edges that pair up the vertices, that is, each vertex is the end of exactly one edge in the perfect matching. If costs are assigned to the edges, then the problem is to find a perfect matching such that the total cost of the edges in the matching is as small as possible. Like the TSP, it is not at all clear how to efficiently find such a matching.

Since we do not know how to optimize, we better call in LP duality. Indeed, perfect matchings satisfy a form of the degree constraints, namely, the edges meeting each vertex must sum to one. This is a good starting point, but cutting planes are needed. And Edmonds delivered. His *blossom inequalities* capture the fact that a graph with an odd number of vertices cannot have a perfect matching. Thus, for any odd cluster of vertices, a perfect matching must contain at least one edge connecting the cluster to the remainder of the graph. The corresponding linear inequality has the form of a subtour-elimination constraint, but this time we ask only that the sum of variables be at least one rather than at least two.

Edmonds adds in the full set of blossom inequalities, one for each odd set S, in a single wave of cutting planes, and proves that the resulting polytope is in fact the convex hull of the perfect matchings of the graph. With such a description, he is able to directly apply LP duality to obtain a polynomial-time algorithm to compute minimum-cost perfect matchings, avoiding a step-by-step cutting-plane process. This is a remarkable result that Edmonds sums up as follows. *"But here is a good algorithm, here is a solved integer program.* And, you know, this was a sermon, this was a real sermon. *Here is a solved integer program.* It was my first glimpse of heaven."[21]

Can this possibly work for the TSP? In 1964, Edmonds discussed the fact that the corner points of the TSP polytope have a simple characterization, despite their huge number, and thus the facets might too have a simple characterization. He writes: "At least we should hope they have, because finding a really good traveling salesman algorithm is undoubtedly equivalent to finding such a characterization."[22] This is a bold statement, but his insight was right on the money. Indeed, the LP algorithm of Khachiyan, that made the *New York Times* front page in 1979, has the interesting property that it can be executed without an explicit list of the constraints of the LP problem, as long as a separation routine is available.[23] Thus, subject to a few technical conditions, Khachiyan's work can be used to show that a good separation algorithm yields a good optimization algorithm, and, vice versa, a good algorithm for optimization yields a good algorithm for separation. This is a deep mathematical result, proven in its most complete form by Martin Grötschel, László Lovász, and Alexander Schrijver in the 1980s.[24]

It follows that to solve the TSP we must have a characterization of the convex hull of tours, equipped with a polynomial-time separation algorithm. Beautiful! The practical, algorithm-engineering side of the TSP calls for faster and better separation algorithms, and the million-dollar Clay Prize is tied to this same strategy. It would be great if this could be worked out, giving another peak at heaven!

Cutting Planes for Integer Programming

Returning to earth, the cutting-plane method is no exception to the rule that successful techniques in integer programming have their roots in TSP research. Cutting planes are far and away the most important tool in modern IP solvers. Subtours, combs, clique trees, et al. are TSP-specific, but the overall method of improving an LP relaxation step-by-step via cutting planes carries over to the general IP setting.

Ralph Gomory was the first person to investigate IP cutting planes in earnest. In later years he would go on to be the senior vice president for science and technology at IBM and the president of the Alfred P. Sloan Foundation, but in the mid-1950s Gomory was tucked away in the mathematics department at Princeton University as a postdoctoral fellow, attempting to squeeze out whole-number solutions from LP problems. Trained in classical mathematics, Gomory sought to apply the theory of Diophantine analysis, that is, the study of integer solutions to linear equations. An extension to the linear inequalities used in LP models seemed promising, but a week of long days and longer nights produced only a set of partially worked-out examples.[25]

> Late in the afternoon of the eighth day of this I had run out of ideas. Yet I still believed that, if I had to, in one-way or another, I would always be able to get at an integer answer to any particular numerical example. At this point I said to myself, suppose you really had to solve some particular problem and get the answer by any means, what would be the first thing you would do? The immediate answer was that as a first step I would solve the linear programming (maximization) problem, if the answer turned out to be 7.14, then I would at least know that the integer maximum could not be more than 7. No sooner had I made this obvious remark to myself than I felt a sudden tingling in two of my left toes, and realized that I had just done something different, and something that certainly was not part of classical Diophantine analysis.

The idea is that if we know an inequality $3x + 2y \leq 7.14$ is satisfied by all solutions to an LP problem, then we know also that $3x + 2y \leq 7$ is satisfied by all integer solutions. Thus, a halfspace that touches the border of the LP feasible region can be pushed in a small amount without cutting off integer solutions.[26]

Gomory worked this observation into an algorithm for solving pure IP problems, that is, LP problems where all variables are required to take on integer values. His method utilizes the form of the simplex algorithm, deriving a cutting plane whenever a dictionary assigns a non-integer value to a variable. Gomory's short paper on the work turned the field of integer programming upside down for a number of years, but once computers became speedy enough to tackle large instances it became apparent that the algorithm did not behave well in practice. Nonetheless, the story has a happy ending: a variant of the mechanism for creating cutting planes, also developed by Gomory, is now the workhorse of commercial IP solvers.

7: Branching

*The basic method will be to break up the set of all tours into smaller
and smaller subsets and to calculate for each of them a lower
bound on the cost of the best tour therein.*

—John Little et al., 1963.[1]

In the Dark Age of TSP cutting planes, between the work of Dantzig
et al. in 1954 and Chvátal in 1973, researchers focused on a variety
of alternative solution methods. Chief among these is the divide-and-
conquer approach known as branch-and-bound, another general-purpose
tool developed first in the context of the salesman. In state-of-the-art TSP
software, branch-and-bound is combined with the cutting-plane method to
produce a powerhouse capable of solving instances of the problem having
thousands of cities.

Breaking Up

The search for an optimal tour hidden in an LP relaxation is a search for
the best needle in a large haystack. With patience and a sufficient supply of
inequalities, the cutting-plane method will eventually solve the problem by
exposing a shiny needle at the top of the stack.

Sounds good, but at some point it may happen that each new cutting
plane removes only a tiny amount of hay. Rather than continue to cut,
cut, cut, we can instead consider splitting the remaining haystack into
two smaller stacks. A good division of this type can shine a light on the
collection of needles, making each of the two newly created subproblems
much easier to solve than a search of the full stack. The splitting step is
known as *branching*; Dantzig et al. described the idea in general terms and
Willard Eastman worked it into a complete TSP algorithm.[2]

We present the method in the next section, but to start let's take a
detailed look at a single branching step for the 42-city USA problem.

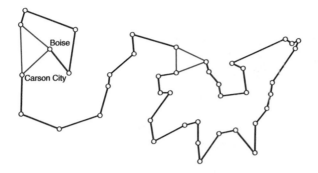

Figure 7.1
Solution after adding
three subtour-elimination
constraints.

Cutting planes alone make short work of this example, so let's handicap the process and permit as cuts only those subtour inequalities that correspond to islands in the LP solution. Such an approach takes us through the first three steps of the computation documented in the previous chapter, resulting in an LP bound of 682.5 units and the connected graph displayed in figure 7.1.

The displayed LP solution sits atop the haystack, so it makes sense to split the problem in such a way that this piece of hay, at least, will disappear. The standard method to accomplish this is quite simple. We select one of the red edges and force the problem to decide to use the edge or not, that is, we split the set of tours into those that do not contain the edge and those that do contain the edge. This works beautifully in the LP model, since we can create the first subproblem by adding the constraint $x_{ij} = 0$ and the second subproblem by adding the constraint $x_{ij} = 1$, where city i and city j are the ends of the red edge. The newly created subproblems are called the 0-side and 1-side of the branch.

In our example we use the connection between Boise and Carson City as a branching edge; the resulting LP solutions are displayed in figure 7.2. Note that the Boise to Carson City edge, indicated by the yellow region, does not appear in the LP solution for the 0-side of the branch, but it does appear in the solution for the 1-side. Again, the simplex method does not lie. The 0-side solution has objective value 687.5 and the 1-side solution has objective value 686. Thus, since each tour through the forty-two cities is a feasible solution to either one or the other of the two LP models, we now know that no tour can have value less than 686 units.

We have increased the 682.5 bound to 686 via the branch, but the situation is even better: the LP solution in the 1-side has islands that can be stamped out with additional subtour-elimination constraints. Doing so results in the new LP solution displayed in the figure, yielding a bound of 703.5. That is good. We know already a tour of length 699, so in searching

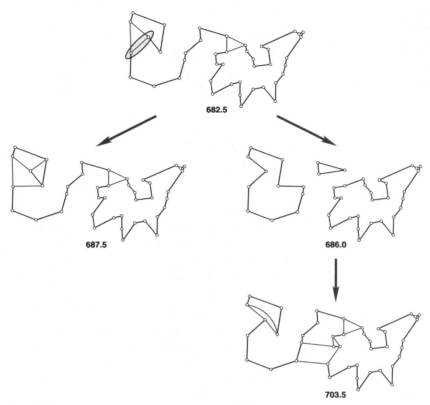

Figure 7.2
Branching on the Boise to Carson City edge.

for a possibly better tour there is no need to look further into this side of the branch. This allows us to *prune* the 1-side subproblem, discarding it from our search.

Summing things up, we split the haystack in two, added a few cutting planes, and then tossed out one of the two resulting stacks. Such a good turn of events is not always possible, but the example should convince you that branching can indeed be a useful tool in a TSP solution process.

The Search Party

The search strategy that grew out of Eastman's work was dubbed *branch-and-bound* by TSP researchers Little et al.[3] The idea is straightforward. We begin with our original problem, which we call the *root* relaxation. If at any point the LP bound associated with a subproblem is greater than or

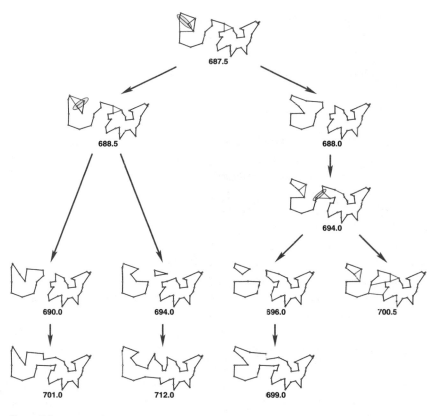

Figure 7.3
Branch-and-bound search tree for the 42-city USA problem.

equal to the length of a known tour, then we prune the subproblem from further consideration. At each step we choose a remaining subproblem and branch, creating two new *child* subproblems. The process stops when each unbranched subproblem has been pruned.

Simple enough, but the method can only succeed if we have a high-quality bounding mechanism. Weak bounds lead to very large search trees, while strong bounds allow for quick pruning and a short path to the optimal solution. Eastman himself did not consider the idea of improving the degree-constraint bounds with cutting planes, and thus he achieved only modest computational success, limited to a ten-city instance worked out in his Ph.D. thesis.

In contrast, the drawings in figure 7.3 show how branch-and-bound solves the 42-city example, starting with the 687.5 model as the root and stamping out subtours whenever they appear in unpruned subproblems.

We have followed Eastman's lead and displayed the search process as a tree, with child subproblems connected to their parent. Note that each branching step leads to children with substantial improvements in the corresponding LP bounds. This is what we like to see.

Branch-and-Cut

Despite an initial rivalry between cutting and branching, the techniques are natural partners. Indeed, the combination of methods swept into power in the 1980s, with Manfred Padberg and Giovanni Rinaldi leading the way. These researchers coined the term *branch-and-cut* and their careful implementation made the process fly for the salesman, smashing previous records with the solution of a 2,392-city test instance.[4]

Padberg and Rinaldi did not handicap the cutting-plane side of their computations. Just the opposite, they unleashed whatever cuts they could obtain in order to improve the LP models. With this approach they found it useful to store their cuts in a *pool* to share among subproblems, that is, cutting planes that improve the LP bound in one subproblem are stored for possible use when processing other subproblems. The purpose of the pool is twofold. First, searching the pool can be much faster than carrying out elaborate separation algorithms. Second, most separation algorithms employed in TSP computations are hit-or-miss heuristic methods. And sometimes they miss at one subproblem but hit at another. By collecting the inequalities into a pool, we gather the hits and see what they can do at other parts of the search tree.

Strong Branching

Vašek Chvátal likes to equate the branching step to marriage. Once we decide to branch we are committed to the new view of the problem; there is no undoing what branching has done. Adopting this metaphor, we can say that early branch-and-bounders typically employed a system of marrying the first potential partner they met on a street, that is, spotting an edge that is assigned a fractional value in the LP solution, they immediately create a pair of subproblems by forcing the edge to go one way or the other. This is a quick method, but in branch-and-cut the bounding process is quite time consuming and it pays to make a greater effort in the selection of a branching edge. Padberg and Rinaldi do this with a technique that we liken to using a dating service to screen potential mates. In Concorde we took this a step further and proposed to first go out on a few dates with a number of candidates before making a selection. Later we went all-in and decided it was best to go ahead and live together with several top choices before committing to the branching marriage.

Let's start with the dating service. Padberg and Rinaldi employed a statistical scheme for building a group of the *most-fractional* edges, that is, edges having value far away from zero and far away from one. From among these, they select the edge having greatest travel cost. This makes a lot of sense. Branching on a long edge will tend to have a greater impact on the LP bounds than branching on a short edge.

The Concorde idea, called *strong branching*, is to take some of the guesswork out of the selection by calling on the LP solver to give a strong hint as to the values that would be obtained by the simplex method. The hint is provided by carrying out a limited number of simplex pivots, say fifty, for each pair of children determined by the branching candidates. With these values in hand, we select the branch that appears to give the greatest increase in the LP bounds. This computation can take considerable time, but it is well worth the effort to avoid a bad split that can potentially double the size of a search tree.

The "all-in" upgrade is to select the top few choices produced by strong branching, then actually solve the corresponding LP subproblems, including a limited application of cutting planes for each potential child. This procedure gives an accurate indication of the bounds we can expect to obtain, but it comes at a great computational cost that pays off only on the most difficult instances. Purists might say we are cheating by trying out branches before making our final selection, but all is fair in the struggle against the salesman.

Branch-and-Bound for Integer Programming

From its TSP roots, branch-and-bound quickly made its way over to general integer programming. The pioneers in this case were Ailsa Land and Alison Doig from the London School of Economics.[5]

In a memoir published in 2010, the two researchers discussed their procedure.[6]

Figure 7.4
Ailsa Land, Banff,
1977.

We did not initially think of the method as "branch and bound", but rather in the "geometrical" interpretation of exploring the convex feasible region defined by the LP constraints. We are not sure if "branch and bound" was already in the literature, but, if so, it had not occurred to us to use that name. We remember Steven Vajda telling us that he had met some French people solving ILPs by "Lawndwa," and realizing that they were applying a French pronunciation to "Land-Doig," so we don't think they knew it as branch and bound either.

The basic Land-Doig process dominated the practical world of IP computation in a way that Gomory's algorithm failed to do, and it was well into the 1990s before branch-and-cut took hold in commercial IP software, finally bringing together the IP rivals.

8: Big Computing

I don't think any of my theoretical results have provided as great a thrill as the sight of the numbers pouring out of the computer on the night Held and I first tested our bounding method.
—Richard Karp, 1985.[1]

The combination of ever-improving mathematical techniques, careful algorithm engineering, and powerful computing platforms has taken the TSP to dizzying heights, but the struggle with the salesman is far from over. Let's see where we stand.

World Records

When it comes to TSP records, nothing comes close to topping the work of Dantzig, Fulkerson, and Johnson.

> It is absolutely astonishing that the three authors were able to find an optimal solution of such a large TSP instance and to prove its optimality by manual computation.
> —George Nemhauser and Martin Grötschel, 2008.[2]

> Dantzig, Fulkerson, and Johnson showed a way to solve large instances of the TSP; all that came afterward is just icing on the cake.
> —David Applegate et al., 1995.[3]

The world eventually understood and digested the work of the RAND team, pushing their techniques in many directions and achieving notable results. But for breadth of new ideas, wide-ranging impact, and quality of exposition, their 1954 paper stands alone as the greatest achievement in the history of the salesman problem.

64 Random Locations

The years immediately following the Dantzig et al. work were quiet ones on the TSP front. Various methods were explored, but computational tests were typically limited to instances of the problem having ten or so cities. Michael Held and Richard Karp finally brought this period of drought to an end in 1971, initiating a big push to larger and more difficult computations.[4]

In his 1985 Turing Award Lecture, Karp stated clearly the intentions of the Held-Karp study. "A few years earlier, George Dantzig, Raymond Fulkerson, and Selmer Johnson at the RAND Corporation, using a mixture of manual and automatic computation, had succeeded in solving a 49-city problem, and we hoped to break their record."[5]

Break it they did, first solving again the 49-city problem, then moving on to a 57-city instance through the United States and a 64-city random Euclidean instance. The algorithm and code that won the day were based on a sophisticated bounding mechanism derived from spanning trees. This bound was incorporated into a branch-and-bound search that produced remarkably small search trees for their test problems.[6]

> It was exciting. We had a lot of trouble and then one day it started working like a charm. The branch-and-bound trees tended to be rather narrow and didn't have many deviations; the bounds were good enough that you went more or less directly to the solution.

The Held-Karp study is unique among the record-setting TSP computations in that it did not directly utilize the cutting-plane method. Their bounding mechanism does, however, have a connection with linear programming and the work of Dantzig et al. Indeed, the Held-Karp bound is an approximation to the optimal objective value of the subtour LP relaxation, and their entire approach can be seen as a means to avoid the use of general LP software. This is not a reflection of the quality of LP solvers at the time, but it was true that available software was difficult to use in the iterative fashion required to run the cutting-plane method.

80 Random Locations

A number of research groups improved on the details of the Held-Karp branch-and-bound process, resulting in the solution of a 67-city Euclidean instance by an Italian team in 1975.[7] But it was inevitable that the cutting-plane method would strike back: the inclusion of inequalities beyond

Figure 8.1
Top: Held and Karp's optimal 57-city tour. Bottom: Michael Held, Richard Shareshian, and Richard Karp, 1964. Courtesy of IBM Corporate Archives.

Figure 8.2
Panagiotis Miliotis, 1974.

subtour-elimination constraints offers too much additional power over bounds that can be achieved with the Held-Karp approach.

The punch landed at the London School of Economics in the group led by Ailsa Land. Her student Panagiotis Miliotis solved a collection of random Euclidean instances having up to 80 cities, utilizing a hybrid of cutting planes and general integer programming, first proposed in the mid-1960s by Glenn Martin.[8] The process runs roughly as follows. Starting with the degree LP relaxation, restrict all variables to take on zero or one values only

Figure 8.3
Martin Grötschel, 2008.
Courtesy of Konrad-
Zuse-Zentrum für
Informationstechnik Berlin.

and apply Gomory's IP cutting-plane algorithm. If the solution is a tour, it must be optimal; otherwise, add a number of violated subtour-elimination constraints, call Gomory's algorithm again, and repeat the process.

The total running time needed by Miliotis to handle the 80-city example was under one minute, suggesting larger instances could also be tackled. Miliotis, however, made the following remark.

> It is unfortunate that the cuts occupy lots of space in the A matrix (the matrix of the coefficients of the constraints). It is characteristic that a 90 cities random Euclidean problem failed with the cutting planes because the non-zero elements in the A matrix exceeded 30,000, most of these occurring in the cutting plane constraints.

This is an unfortunate feature of Gomory's method: each additional cutting plane involves nearly all of the variables in an LP model, which eventually slows the simplex algorithm to such an extent that further progress is impossible. In contrast, the beauty of the pure Dantzig et al. approach is that TSP-specific inequalities tend to assign most edges the value zero, resulting in LP models that are relatively easy to solve.

120 Cities in Germany

Martin Grötschel was the next to step up with a new TSP record. His approach was pure cutting planes: he used LP software to solve his models, but found his inequalities with by-hand calculations, much in the spirit of the 1954 study. We have already seen his optimal 120-city tour of Germany in the three-tours map displayed in figure 1.9 in chapter 1.

In the record-setting computation, Grötschel solved thirteen LP relaxations, using a total of 36 subtour-elimination constraints and 60 comb inequalities. Each of the thirteen rounds took between thirty minutes and three hours of by-hand computations, and between thirty seconds and two minutes of computer time.[9] His LP solver was the IBM package called MPSX. Grötschel described the overall process as follows in an e-mail message from June 11, 2005.

> After an MPSX-run I printed the solution and drew the corresponding solution picture. Then I made some copies of the solution picture and tried to detect cutting planes. Of course, after having gained some experience, I could find lots of violated inequalities. But in those days MPSX was not able to handle very large scale linear programs so that I restricted myself to adding something like 5 to 20 cuts per run making educated choices as to which of the cutting planes that I spotted would do a reasonable job.

It was a great achievement, demonstrating the power of comb inequalities in solving large-scale instances of the TSP.

318 Holes in a Circuit Board

Hot on the heels of Grötschel's 120-city computation, Manfred Padberg initiated a joint study with Saman Hong, a recent Ph.D. graduate from Johns Hopkins University. Their project was a computational success, automating the cutting-plane algorithm, solving instances with up to 75 cities, and computing good lower bounds on other instances.[10] The largest example treated in the study was a 318-city drilling problem considered earlier by Shen Lin and Brian Kernighan.

Not satisfied with good approximations, Padberg continued his pursuit for a solution to the Lin-Kernighan example several years later, this time together with IBM's Harlan Crowder.[11]

> One evening we had it all together and submitted a computer run for the 318-city symmetric TSP. We figured it would take hours to solve and went to the "Side Door", a restaurant not far from IBM Research, to have dinner. On the way back we discussed all kinds of "bells and whistles" we might want to add to the program in case of a failure. When we got to IBM Research and checked the Computer Room for output it was there. The program proclaimed optimality of the solution it had found in under 6 minutes of computation time!

The Crowder-Padberg study concluded with the solution to the 318-city instance and a large collection of smaller examples.[12]

Figure 8.4
Left: Optimal solution of a 318-city drilling problem. Right: Manfred Padberg (second
from right) and Harlan Crowder (sitting), 1982. Courtesy of Manfred Padberg.

666 World Locations

This brings us to 1987, a high point in TSP computation rivaled only
by the Big Bang of 1954. In the first of two major studies announced
that year, Martin Grötschel and Olaf Holland tackled a large set of test
instances, the most difficult of which consisted of 666 cities chosen from
around the world. Bible scholars will recognize 666 as "the number of the
beast," and, indeed, Grötschel selected the cities with the idea of creating
a beastly challenge for TSP computations.[13] Grötschel and Holland met
the challenge with a hybrid of the cutting-plane method and general
integer programming, this time with IBM's MPSX-MIP/370 code as the
IP solver. Their computational success centered around a host of new exact
and heuristic separation algorithms for comb inequalities, allowing their
code to obtain very strong LP relaxations.

Figure 8.5
Left: Grötschel and Holland's 666-city tour. Right: Olaf Holland, 2010.

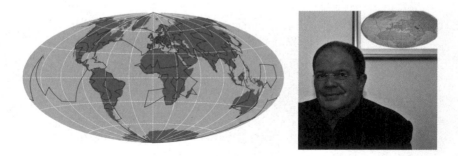

2,392 Holes in a Circuit Board

The hybrid approaches adopted in work by Glen Martin, by Panagiotis Miliotis, and by Grötschel and Holland permit the use of high-powered IP machinery, but the second major study of 1987, by Manfred Padberg and Giovanni Rinaldi, demonstrated clearly the benefits of keeping the solution process squarely in the TSP arena. Branch-and-cut was the method wielded by Padberg and Rinaldi. Their computer code solved test instances that included a 532-city USA tour, the Grötschel-Holland 666-city challenge problem, and drilling problems with 1,002 cities and 2,392 cities. The solution of the 2,392-city TSP was a stunning accomplishment and easily the most complex attack on an optimization problem up to that time.[14]

The high-profile Padberg-Rinaldi study brought in major computing hardware, including a CDC Cyber 205 supercomputer at the National Bureau of Standards and the vector-processing IBM 3090/600 computer at the IBM T. J. Watson Research Center. But it was the algorithmic work that pushed the TSP results forward, including new separation heuristics for combs and clique trees and numerous innovations in implementing branch-and-cut. Their paper concludes with the optimistic note, "the problem with 2392 cities should not be the end of the ongoing saga of the symmetric traveling salesman."

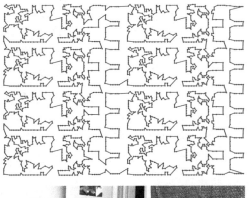

Figure 8.6
Top: Optimal tour of a 2,392-hole printed circuit board. Bottom: Manfred Padberg and Giovanni Rinaldi, 1985. Courtesy of Manfred Padberg.

3,038 Holes in a Circuit Board

Dave Applegate, Bob Bixby, Vašek Chvátal, and I began our joint work in 1988, hoping indeed to continue the saga. We initially attempted to avoid the cutting-plane method, adopting the motto that if other people have tried something then it was not for us. It didn't take long to realize this was a big mistake and by 1989 we were immersed in the details of branch-and-cut. Our initial code was called "Subtour," since it was first created only as a means to compute the optimal solution of the subtour LP relaxation, which we used as a measure of the performance of our non-cutting-plane bounds. It eventually grew to include an array of new separation routines, as well as the strong-branching methods we described earlier.

The home for our computational work was the Bellcore research lab in New Jersey, which housed fifty or so desktop workstations that could be harnessed when their owners were away from their offices. These machines on their own were not comparable to the big-iron hardware employed in the Grötschel-Holland and Padberg-Rinaldi studies, but working together on a network they could be a speed demon. It was thus natural to develop a parallel-processing approach, dividing up the work of finding cutting planes and processing subproblems.

Unfortunately we were not the only ones hunting for free computer cycles. The rival was Arjen Lenstra, who led the computation to factor the 129-digit RSA challenge number into its prime components.[15] The contest for computing time was not entirely friendly, but in the end both teams came away with big slices of the available hardware. Our strategy was simple: we only started our software on a machine if it was currently free, other than the pesky factoring code run by Lenstra, and we terminated the code as soon as we sniffed the owner returning to his or her desk. Dave Applegate implemented this strategy with a small code that checked for user input, such as a mouse movement or a keyboard stroke, several times a second. Dave's code made it extremely difficult for a user to even know that we had taken over his or her machine: if the user issued a command to see what was running, then our code disappeared before the command's results could be delivered to the screen.

Using this Bellcore network, our first result came in 1992 with the solution of a 3,038-city circuit-board drilling problem from Gerd Reinelt's TSPLIB. With various tweaks to the code and algorithms, we were able to move up to a 4,461-city tour through the old East Germany in 1993, and a 7,397-city solution to a computer-circuit problem in 1994. At this point we agreed to conclude our TSP project, but, perhaps fortunately, the wrapping-up process did not work out as planned.

Figure 8.7
Left: David Applegate.
Right: Robert Bixby.
Photographs by
Jakob Schelbert,
Erlangen, 2010.

13,509 Cities in the United States

With the conclusion of our computations, we set out to document the techniques employed in the run up to the 7,397-city TSP. Here we ran into a problem. It would be difficult to convince other researchers that we had developed a solid approach when we ourselves were not happy with the details of the work; the Subtour code had grown in spurts as we discovered new techniques and it did not capture well our global view of TSP computation. Faced with this fact, we did the only sensible thing and threw it away. Rather than writing up documentation, we began work from scratch on a new code than we called "Concorde."

This restart of the project was a luxury and we took advantage of the opportunity to include algorithms and techniques aimed at much larger TSP instances. One of the key new components was a "local cuts" separation routine based on shrinking clusters of cities to obtain very small graphs, allowing us to apply a time-consuming LP-based method to obtain cutting planes.

The main target of our computational study was a 13,509-city tour through the United States that was solved in 1998. In the years following this result we continued to work, but it was mainly work to understand what we already had in place, similar to a race car driver getting used to a new vehicle. Along the way we computed optimal tours for 15,112 cities in Germany in 2001 and 24,979 cities in Sweden in 2004.

85,900 Gates on a Computer Chip

The Sweden TSP computation pushed the Concorde code to its limit. The work was carried out on a cluster of 96 dual-processor computers at Georgia Tech, running as a background process when the machines were not otherwise active. The total amount of computing time was a whopping 84.8 years.

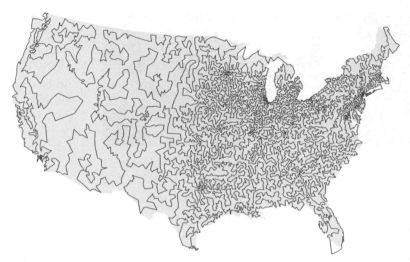

Figure 8.8
A tour of all 13,509 USA cities having population of at least 500.

Figure 8.9
Left: Daniel Espinoza.
Right: Marcos Goycoolea.

We were quite proud of the Sweden tour, but the TSPLIB contained two larger instances, having 33,810 cities and 85,900 cities respectively, and these appeared to be beyond the reach of Concorde. Fortunately, Daniel Espinoza and Marcos Goycoolea's implementation of Letchford's separation algorithm came online at about this point (see the discussion at the end of the separation problem section in chapter 6), giving us enough horsepower to take a shot at completing the solution of Reinelt's TSPLIB collection.

The final run on the 85,900-city TSP was begun in February 2005 and it concluded with an optimal tour in April 2006. A plot of the steadily rising LP bound is given in figure 8.10; the data points are taken from the nearly daily logs of the computations, with a break in December 2005 when the cluster of machines was down for repair. The bound eventually grew to

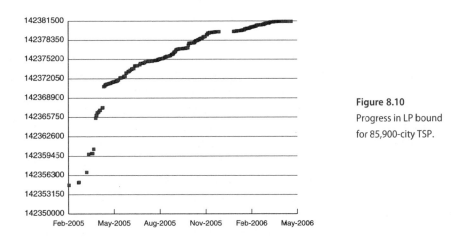

Figure 8.10
Progress in LP bound
for 85,900-city TSP.

within 0.001% below the length of the best-known tour found in 2004 by Keld Helsgaun. This was close enough for a short branch-and-cut search to close the remaining gap and prove that Helsgaun's tour was indeed optimal.

The 136 years of computing time used in the calculations made it difficult, of course, for anyone to verify our claim that we had actually solved the problem. Thus, in keeping with its record status, in 2009 we published a computer code and data set that together certify the optimality of the 85,900-city tour.[16] The data consists of the cutting planes and dual LP solutions for each subproblem in the branch-and-cut search. The computer code, a relatively svelte 6,646 lines in the C programming language, runs through the subproblems and establishes that the dual LP solutions provide bounds to allow each of them to be pruned. Not as clean, perhaps, as a proof of the Pythagorean theorem, but it does set down enough information for future TSP researchers to dig through the results of the huge computation.

The TSP on a Grand Scale

The beauty of the computational side of the TSP is the simple fact that there is always a larger problem.

Bosch's Art Collection

Moving just beyond the TSPLIB test collection, I really like the challenge instances created by Bob Bosch, using a technique we describe in chapter 11. His six problems range in size from the 100,000-city Mona Lisa up to a 200,000-city rendering of Vermeer's *Girl with a Pearl Earring*. The data sets are on the Web, available to anyone wanting to take a crack at finding better tours or establishing bounds on possible tour lengths.[17]

Among Bosch's test problems, the Mona Lisa TSP has attracted by far the most attention, sitting as it does just tantalizingly above the 85,900-city record solution. The small difference between 85,900 cities and 100,000 cities may be misleading, however. It seems likely that by almost any measure the Mona Lisa TSP is much more challenging than the current world record computer-circuit example. Indeed, the image displayed in figure 1.8 reveals many cities close together on straight lines. This geometry suggests that the 85,900-city instance may be somewhat easy relative to its size, whereas the distribution of points in the Mona Lisa TSP looks as complicated as one might imagine.

We mentioned in chapter 1 that there is a $1,000 bounty for a Mona Lisa tour shorter than the current best found by Yuichi Nagata on March 17, 2009. Nagata's computation culminated a flurry of activity among the world's top tour finders in February and March of 2009, just after Bosch created his data sets. The record for the best solution changed hands six times during this period, with Nagata's tour coming in at length 5,757,191, an improvement of eight units over a tour found the previous day by Keld Helsgaun. But is Nagata's tour optimal?

On January 18, 2010, a bound of 5,757,044 was established for the Mona Lisa TSP, leaving a gap of only 147 units to Nagata. This seems tiny, indeed only 0.0026%, but a gap is still a gap. The bound was obtained via a Concorde branch-and-cut search having 1,065 subproblems, run over the course of 66 days and 4.37 years of computer time. Further runs with Concorde could probably push the gap down a few more units, but to conclude things we almost certainly need new ideas, particularly in cutting-plane separation. For computation nuts like me, this is where the fun begins.

The World

The World TSP challenge awaits anyone with ideas for big-time improvements in TSP computation. The data for the 1,904,711-city instance was obtained from the *National Imagery and Mapping Agency* and the *Geographic Names Information System*. At the time of its creation in 2001, the problem covered every point on the globe populated by humans. The points are specified by their latitude and longitude, and the cost of travel between pairs is given by an approximation of the great circle distance, treating the earth as a ball; this cost function is a variation of the TSPLIB GEO-norm, scaled to provide distances in meters rather than in kilometers.

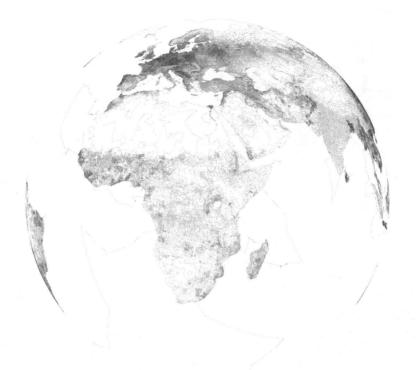

Figure 8.11
A tour of every city in the world. Image courtesy of David Applegate.

An initial tour of length 7,539,742,312 meters and a 7,504,218,236 bound established an optimality gap of 0.47% in the fall of 2001. This has been chipped away steadily over the past decade, primarily by Keld Helsgaun on the tour side and Concorde on the bound side. A chart of the progress in the gap is displayed in figure 8.12; the red bars indicate improvements through better bounds and the grey bars indicate improvements through better tours. The current status is Helsgaun's tour of length 7,515,790,345 and Concorde's LP bound of 7,512,218,268, yielding a gap of 0.0476%.

The narrowing of the gap over the past decade to just over one-tenth of its original value is nice progress. An interesting fact is that nearly 75% of this improvement was achieved through better tours. The TSP world in 2001 would not have thought this to be possible: the commonly held view was that state-of-the-art TSP heuristics produced near-optimal tours

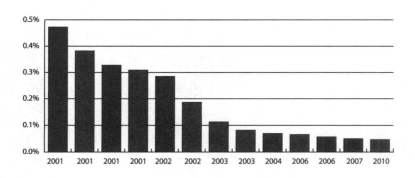

Figure 8.12
Decreasing optimality gap for World TSP.

and thus the initial optimality gap must be due to a weak LP bound. Helsgaun's LKH code changed this perception, demonstrating that there is still room for improvement in tour-finding methods. At this point I would not be able to guess where the optimal value lies, either closer to the tour of 7,515,796,609 or closer to the bound of 7,512,218,268. It is certainly true, however, that the LP bound can be raised through engineering improvements in Concorde. For example, currently it is not possible to employ the domino-parity-separation module due to the size of the data set. This is good news. More work to do!

The Stars

As perhaps a final stop, at least for the foreseeable future, Dave Applegate and I created in 2003 a TSP instance for the 526,280,881 celestial objects in the United States Naval Observatory A2.0 catalog. We initially wanted to include estimates on distances between pairs of objects, thinking it would be fun to send the *Enterprise* on an optimal tour in its next five-year mission. But the coarseness of the data would place all stars on a small number of concentric spheres, so we opted instead to model the movement of a telescope. Thus the data specifies cities as locations in the sky, with travel costs given by a measure of the angle determined by each pair of cities.

The Star TSP is very difficult to handle in one piece; even an n^2 running time is enormous when n is five hundred million. Thus the goal for now is to develop methods that split the data set apart, establish bounds and tours in the components, and then piece things back together. Dave Applegate,

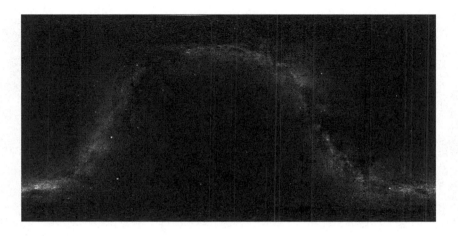

Figure 8.13
Skymap. Image courtesy of NASA/Goddard Space Flight Center Scientific Visualization Studio.

Keld Helsgaun, Andre Rohe, and I made initial experiments with this approach, establishing in 2007 an optimality gap of 0.410%. This puts the problem about ten years behind the World TSP, so if you have wild ideas for tours or bounds, please keep this big data set in mind.

9: Complexity

*Is there or is there not a polynomial-time algorithm for obviously
finite bounded integer programming? I mean, this is my sermon:
Is there or is there not?*

—Jack Edmonds, 1991.[1]

The pursuit of the salesman through ever larger numbers of cities has led to breakthroughs in mathematics, computing, and engineering, as well as advances in numerous practical applications. This is the pride and joy of TSP researchers. But the step-by-step approach has not answered the mother-of-all complexity questions: can we efficiently solve *every* instance of the TSP?

The fate of the salesman from this complexity point of view is tied to that of many other problems, including general integer programming, via the theory of Stephen Cook and Richard Karp. Indeed, the TSP is rolled into the $\mathcal{P}$ vs. $\mathcal{NP}$ question, one of the seven *Millennium Problems* for which the Clay Mathematics Institute offers a $1,000,000 prize. The Clay Web site introduces the challenge as follows.

> If it is easy to check that a solution to a problem is correct, is it also easy to solve the problem? This is the essence of the $\mathcal{P}$ vs. $\mathcal{NP}$ question. Typical of the $\mathcal{NP}$ problems is that of the Hamiltonian Path Problem: given N cities to visit (by car), how can one do this without visiting a city twice? If you give me a solution, I can easily check that it is correct. But I cannot so easily (given the methods I know) find a solution.

This is typical of public descriptions of $\mathcal{P}$ vs. $\mathcal{NP}$, where the TSP or one of its variants is used to motivate the question. In this chapter we discuss what we know and do not know about the general complexity of the salesman.

A Model of Computation

To have meaning, mathematical statements must be precise, or at least it must be true that they can be made precise. The million-dollar complexity question is no exception. Here we need to be precise in what we mean by an algorithm, that is, what does it mean to be computable? This issue came to the forefront in the early 1900s with David Hilbert's *Entscheidungsproblem*, that asks, roughly, whether there exists an algorithm that can decide if any given statement is, or is not, provable from a set of axioms. The development of theory to handle such questions is a beautiful achievement of twentieth-century mathematics, with giants Kurt Gödel, Alonzo Church, and Alan Turing leading the way.

The intuitive concept of an algorithm is that of a list of simple steps that together produce a solution to a problem. Euclid gave us an algorithm for greatest common divisors some 2,300 years ago, but at the time of Hilbert it was not clear how an algorithm should in general be defined. In his famous paper from 1936, Turing provided an answer, introducing a mathematical model known as a *Turing machine*.[2]

Turing's machine has a tape for holding symbols, a head that moves along the tape reading and writing symbols in individual cells, and a controller to guide the read/write head. The machine also has a finite set of states, with two special states being *initial* and *halt*. The controller is actually a table that indicates what the machine should do if it is in a particular state s and it reads a particular symbol x. The "what it should do" is to print a new symbol x' on the cell of the tape, move the head either left or right one cell, and enter a new state s'. To solve a problem, the machine starts in its *initial* state, with the input to the problem written on the tape; it terminates when it reaches the *halt* state.

It is fun to think about a physical version of a Turing machine, with a read/write head zipping around a long, thin tape full of symbols. Indeed, there are photographs on the Web of students gathered around modified shoe boxes with rolls of tape sticking out the ends, recording transitions from one state to the next on a scratch pad. Turing himself does not mention physical machines, but rather emphasizes, by means of examples in his paper, the fact that a machine can be fully described by writing down its table of transitions from one state to another.

Let's consider a simple case: given a string of 0's and 1's, determine if the number of 1's is odd or even. To construct a Turing machine for this problem we can have four states, *initial*, *odd*, *even*, and *halt*, two symbols, 0 and 1, and the transition table displayed in figure 9.1. The table has a

	initial	odd	even
0	_, right, *even*	_, right, *odd*	_, right, *even*
1	_, right, *odd*	_, right, *even*	_, right, *odd*
_	0, _, *halt*	1, _, *halt*	0, _, *halt*

Figure 9.1
Transition table for a parity-checking Turing machine.

row for each symbol (including a blank "_") and a column for each state
other than *halt*. The entry in the table is a triple, giving the symbol to
write, the direction to move on the tape, and the next state. For example,
if the machine is in state *odd* and we read the symbol 1, then we write a
blank symbol in the cell, move one cell to the right, and change the state to
even. Presented with a string of 0's and 1's, arranged in consecutive cells on
the tape, and with the read/write head positioned on the leftmost symbol,
our Turing machine will move to the right until it reaches a blank cell,
indicating the end of the string. When it halts, a 0 is written on the tape
if the number of 1's is even, and a 1 is written on the tape if the number of
1's is odd. Simple enough, but it illustrates the idea of operating via a table
of transitions.

A next step up is to build a Turing machine that can add two numbers
given in binary notation. This is a common exercise in university-level
courses in computational complexity, and it is kind of fun, as far as such
exercises can be. If you want to get a better feeling for the operation of
a machine, I recommend you give it a try. One thing you will notice is
that it would be convenient to have a second tape to use for intermediate
calculations. Such multiple-tape Turing machines are defined in a natural
way, with separate read/write heads for each tape. Although extra tapes are
convenient, anything we can compute on a multiple-tape machine can also
be computed on a machine with a single tape, albeit a bit more slowly.

This last point, about simulating a multiple-tape machine with a single
tape, is important. We would like to define an algorithm as something that
can be carried out on a single-tape Turing machine, but does this capture
everything we might want in an algorithm? All we can say is that, so far,
Turing machines have been able to handle everything that has been thrown
at them. If something is computable on a modern-day computer, then a
lightning-fast Turing machine could also carry out the computation.

The working assumption that we can equate algorithms and Turing
machines is known as the *Church-Turing Thesis*.[3] This thesis is widely

accepted and it gives the formal model of an algorithm that is used to make precise $\mathcal{P}$ vs. $\mathcal{NP}$ and other complexity questions. Perhaps, one day, exotic computing capabilities will come along to cause us to consider an expanded definition of an algorithm, but for over seventy years Turing has served up just what the research community has needed.

Universal Turing Machines

There is a fundamental difference between a modern telephone, such as the *iPhone*, and, say, a pair of shoes. Shoes are designed for the single duty of protecting your feet, whereas the hundreds of thousands of applications available for the iPhone allow it to take on tasks that were not imagined when the hardware was designed. This is something we take for granted, but the creation of programmable machines was an intellectual leap, made by Turing in his original paper.

A Turing machine is a great model for describing what we mean by an algorithm, but a Turing machine is designed for a single task, such as adding two numbers. In this sense, Turing machines are closer to a pair of shoes than to an iPhone. A crucial point made by Turing, however, is that one can design a *Universal Turing machine* capable of simulating every Turing machine.

> It is possible to invent a single machine which can be used to compute any computable sequence. If this machine $\mathcal{U}$ is supplied with a tape on the beginning of which is written the S.D of some computing machine $\mathcal{M}$, then $\mathcal{U}$ will compute the same sequence as $\mathcal{M}$.

The "S.D" in the quote is short for "standard description," Turing's name for a transition table. The concept is thus to include the table as part of the input on the tape, just like we include a program on a modern computer. Turing's idea, and plenty of hard work by Konrad Zuse, John von Neumann, and others, ushered in the age of computing.

The Campaign of Jack Edmonds

Turing answered brilliantly David Hilbert's call for a theory of algorithms. Once digital computers started to appear, however, the issue of efficiency soon became of fundamental importance. It is one thing to know a problem can be solved by a Turing machine, it is another altogether to know the Turing machine will deliver its solution during our lifetime.

Early discussions on the efficiency of algorithms centered around the TSP and other integer-programming models. A typical quote from this period is the following from Martin Beckmann and Nobel Prize winner Tjalling Koopmans, taken from a 1953 research paper.[4]

It should be added that in all the assignment problems discussed, there is, of course, the obvious brute force method of enumerating all assignments, evaluating the maximand at each of these, and selecting the assignment giving the highest value. This is too costly in most cases of practical importance, and by a method of solution we have meant a procedure that reduces the computational work to manageable proportions in a wider class of cases.

Beckmann and Koopmans considered a family of problems, including both the TSP and the standard assignment problem of matching workers to jobs. The following year, Merrill Flood made a case for efficient solution methods.[5]

There is a rumor that the Navy is building a computer to handle various versions of the tanker scheduling problem. The important thing gained by this would not be economy in carrying out the calculations but rather a shortening of the length of time it would take to recompute the operation. I can't stress this point too much ...use of a high speed computer may indeed cost more dollars yet make the calculation possible in time.

With this need-for-speed in mind, Flood goes on to comment that there "are as yet no acceptable computation methods" for the TSP. So finite is not good enough, but what should be our target when judging the quality of an algorithm? Despite the great need, no clear concept emerged in the 1950s.

This is where Jack Edmonds came onto the scene. We use the word "campaign" in the title of this section, and we mean this literally—Edmonds had to fight against a consensus that better-than-finite was not something to be handled by the mathematics community. Flood himself made the following remark in 1954. "The problem of finding a solution that is practical for use with available computing machinery is, to the pure mathematician, not normally an interesting problem." This was precisely the difficulty. It is true that Flood, Koopmans, Kuhn, and others were interested in practical solution techniques, but here the amazing success of the simplex method for linear programming may have held back direct discussions of better-than-finite algorithms. The troublesome point was that the simplex method

Figure 9.2
National Bureau of Standards workshop, 1964; Jack Edmonds on far right.
Photograph courtesy of William Pulleyblank.

appeared to solve every linear-programming problem in sight, although it could not be proven to always run efficiently. This led to an over-comfortable acceptance of algorithms whose performance could not be guaranteed.

Edmonds had his work cut out for him. His efforts of persuasion began at the RAND Corporation in the summer of 1961, where he joined a group of young researchers invited to take part in a workshop together with leading figures such as Dantzig, Fulkerson, Hoffman, and others. Edmonds's RAND lecture concerned the problem of finding optimal matchings in a graph. During the course of the workshop he succeeded in producing a *good algorithm* for the problem, that is, an algorithm whose number of steps grows at a rate at most proportional to n^4, where n is the number of vertices of the graph. This deep result served as the focal point of Edmonds's campaign, and the beauty of his mathematics helped sway opinions. But it was not without a series of trials for Edmonds, who writes the following in a highly recommended memoir.[6]

> The reactions I would get when I was ranting about this at the time—I remember my obsessions and my talking at full tilt—the biggest reaction I got is: "Well, it's kind of silly to expect such a thing, and let us see, it doesn't have any real meaning, oh, and so what, if it were n to the 28th, you know, that doesn't ...", and all that kind of stuff.

No one questions his theory today. Edmonds is the hero of algorithms and computational complexity.

One thing that did not stick is the use of the word "good" to describe an algorithm that comes with a guarantee to complete its work in time at most proportional to n^k, where n is a measure of the size of the problem and k is some fixed power. As we mentioned in chapter 1, the standard term is now *polynomial-time algorithm*. This was probably a change for the better, since it is a bit harsh to say an algorithm as successful as the simplex method is bad.

Cook's Theorem and Karp's List

Things moved quickly in the early days of computational complexity. In 1967, fresh from successes with matchings and other combinatorial problems, Edmonds turned things upside down by conjecturing that there will never be a good algorithm for the TSP. Why would he expect his polyhedral methods to fail on the salesman, after working spectacularly in other cases? Edmonds was coy with his explanation, noting only that it was a legitimate possibility that no good algorithm exists. Four years after this bet, Stephen Cook and Richard Karp developed their theory placing the question in the larger world of $\mathcal{P}$ vs. $\mathcal{NP}$.

The Complexity Classes

Mathematicians like to keep things tidy, and in the case of complexity theory this has led to a focus on *decision problems*, that is, a focus on problems having yes or no answers. So, for example, does a graph have a Hamiltonian circuit? Yes or no. Or, given a set of cities, is there a tour of length less than 1,000 miles? Yes or no.[7]

Among decision problems, Richard Karp introduced the short notation $\mathcal{P}$ to denote those that have good algorithms. Formally, $\mathcal{P}$ is the class of problems that can be solved in polynomial time on a single-tape Turing machine, that is, if n is the number symbols on the input tape, then the machine is guaranteed to halt after a number of steps that is at most C times n^k, for some power k and some constant C. This definition of $\mathcal{P}$ is certainly tidy: we could replace the single-tape Turing machine by a multiple-tape machine, or even by a powerful modern digital computer, without altering the class. Indeed, the simulation of a modern computer via a Turing machine will slow down computations, but the slow-down factor is only polynomial in n. So if we have a polynomial-time algorithm on a modern computer, then we also have a polynomial-time algorithm on a single-tape Turing machine.

Belonging to $\mathcal{P}$ is the gold standard for decision problems, but Stephen Cook studied a possibly larger class that arises in a natural way. Encapsulating a notion of Edmonds, Cook considered the problems such that yes answers can be verified in polynomial time. To verify an answer, we provide a certificate together with the statement of the problem, allowing a Turing machine to check that the answer is indeed yes. For example, to verify that a set of cities can be visited in less than 1,000 miles, we can provide the machine with such a tour.[8]

An alternative view of verifications is via *nondeterministic Turing machines*. Such "machines" are not part of the physical world, since they have the capability of duplicating themselves during a computation. If there is a polynomial-time verification, then a nondeterministic machine can guess the correct certificate in one of its many copies and determine that the answer to the problem is yes. This view led Karp to propose the shorthand $\mathcal{NP}$ for Cook's class of problems.

On the surface, it would appear to be much easier to be a member of $\mathcal{NP}$ than to be a member of $\mathcal{P}$. The TSP is a case where checking a solution is easy, but finding the solution may be difficult. As a second example, consider the factoring problem of writing a whole number as the product of two smaller whole numbers. This may be a difficult task (no polynomial-time algorithm is known), but it is a simple matter to check that an answer is correct. Many more examples can be constructed, but these only hint that $\mathcal{NP}$ is a larger class than $\mathcal{P}$. An amazing fact is that we do not know of a problem that is in $\mathcal{NP}$ but definitely not in $\mathcal{P}$.

Reducing One Problem to Another

In the paper that began the formal study of $\mathcal{NP}$, Stephen Cook put forth a certain problem in logic as a candidate for a member that may itself not be in $\mathcal{P}$. "Furthermore, the theorems suggest that {*tautologies*} is a good candidate for an interesting set not in $\mathcal{L}^*$ and I feel it is worth spending considerable effort trying to prove this conjecture. Such a proof would be a major breakthrough in complexity theory."[9] Indeed, a proof of his conjecture is now worth a million dollars. Cook's paper predates Karp's introduction of the now standard terminology—his $\mathcal{L}^*$ is what we call $\mathcal{P}$ and his {*tautologies*} is commonly known as the *satisfiability problem*. This problem is to determine whether or not true and false values can be assigned to a collection of logical variables so as to make a given formula evaluate to true. The components of the formula are the variables and their negations, joined up by logical *and*'s and logical *or*'s. More important than the problem itself is Cook's reason for making the conjecture: his theorems show that every problem in $\mathcal{NP}$ can be formulated as a satisfiability problem.

The key component of Cook's theory is the idea of reducing one problem to another. We have seen this in practice several times in the book. For example, Karl Menger's Vienna colloquium presented the problem of finding a shortest path through a set of points. This is not exactly the TSP, since we are not required to return to the starting point. But if we know how to solve the TSP, then we can solve Menger's problem by adding a dummy city with zero travel costs between the dummy and each of the original points.

Formally, a *problem reduction* is defined as a polynomial-time Turing machine that takes any instance of problem A and creates an instance of problem B, such that the answers to A and B are the same, either both yes or both no. In reducing Menger's problem to the TSP, we add one extra city and n extra distances, so we can build a Turing machine to carry out the reduction in a number of steps proportional to n, the number of points in the problem. This is typical of problem reductions, where all you really need to keep in mind is that the size of problem B should not be too much larger than the size of problem A.

It is clear that reductions are useful in sorting out the many members of $\mathcal{NP}$. To show a problem is easy, you can try to reduce it to another easy problem. To show a problem is hard, you can try to reduce a known hard problem to your problem. But the amount of order provided by reductions is surprising: Cook proved that every problem in $\mathcal{NP}$ can be reduced to the satisfiability problem.

The notion that there could be one problem to rule them all is exceptionally deep thinking, but the proof of Cook's theorem turns out not to be that difficult. You can start a proof by noting that since there will be only polynomially many Turing-machine steps in a verification of an $\mathcal{NP}$ problem, to reduce such a problem to satisfiability we can include logical variables for every step of the verification, indicating the state of the machine and the symbol that is read. We will not go into the details here, but we note that the full proof in Cook's original paper occupies less than a page (albeit in a rather small typeface).

Now we can understand Cook's reasoning. A problem reduction from A to B implies that if B is in $\mathcal{P}$, then so is A. Thus, if satisfiability is in $\mathcal{P}$, then there exist polynomial-time algorithms for every problem in $\mathcal{NP}$. Cook thought it unlikely that $\mathcal{P} = \mathcal{NP}$, hence the conjecture.

Twenty-one $\mathcal{NP}$-complete Problems

Calling satisfiability the "one problem to rule them all" fits Cook's result, but it does not capture the full import of his theory of problem reductions. Indeed, knowing satisfiability rules all of $\mathcal{NP}$ gives a direct route for showing that other problems too have this property.

An $\mathcal{NP}$ problem is called $\mathcal{NP}$-complete if every member of $\mathcal{NP}$ can be reduced to it. Cook followed his proof that satisfiability is $\mathcal{NP}$-complete with a quick argument that a graph-theory problem known as subgraph isomorphism is also $\mathcal{NP}$-complete: he showed that satisfiability can be reduced to subgraph isomorphism. So, any member of $\mathcal{NP}$ can be first reduced to satisfiability and then reduced to subgraph isomorphism. Building a single Turing machine to carry out both problem reductions, one after the other, shows that subgraph isomorphism is $\mathcal{NP}$-complete.

This idea of chaining together problem reductions created an explosion of interest in complexity theory, led by a research paper of Richard Karp, written one year after the announcement of Cook's results.[10] Karp gives a delightful technical exposition of $\mathcal{P}$, $\mathcal{NP}$, Turing machines, and reductions. His paper also presents a now famous list of twenty-one $\mathcal{NP}$-complete problems, together with their reductions from Cook's satisfiability problem. The list includes two versions of the TSP: the Hamiltonian-circuit problem for undirected graphs and the Hamiltonian-circuit problem for directed graphs.

Once Karp's paper hit the streets, reductions to other difficult problems followed left and right. Hundreds of problems were shown to be $\mathcal{NP}$-complete, and in 1979 Michael Garey and David Johnson published a landmark book titled *Computers and Intractability: A Guide to the Theory of $\mathcal{NP}$-Completeness*. Their volume lives on the shelf of nearly everyone working in algorithms; when presented with a new problem, one first scans the Garey-Johnson catalog of $\mathcal{NP}$-complete members for likely candidates to use in a problem reduction.[11]

A Million Dollars

As a practical matter, once a problem is shown to be $\mathcal{NP}$-complete, researchers assume it will be nasty to solve and turn either to quick-and-dirty heuristic methods or to one of the heavy-duty approaches pioneered by attacks on the TSP. The working hypothesis is that there can be no efficient polynomial-time algorithm for an $\mathcal{NP}$-complete problem. But there is no compelling evidence that $\mathcal{P}$ and $\mathcal{NP}$ are actually distinct. So, which side do you cheer for? $\mathcal{P} = \mathcal{NP}$ or $\mathcal{P} \neq \mathcal{NP}$?

The ever-growing influence of computation has made the $\mathcal{P}$ vs. $\mathcal{NP}$ question perhaps the most prominent open problem in all of mathematics. Although there has been no shortage of attempts to settle the issue, Lance Fortnow, writing on its status in 2009, summed up his article with two words: "Still open." Nonetheless, with the big $1,000,000 Clay Prize looming, one can hope for progress on the horizon. As Douglas Adams's character Wowbagger the Infinitely Prolonged from *The Hitchhiker's Guide to the Galaxy* said when questioned about the feasibility of his own TSP project of insulting every man and woman in the universe: "A man can dream, can't he?"

State of the TSP

Eindhoven University of Technology's Gerhard Woeginger is the unofficial archivist of numerous brave claims for the Clay Prize. The highlight of his $\mathcal{P}$ vs. $\mathcal{NP}$ Web page is a chronological list of milestones, tagged as "Equal" or "Not equal" according to the purported result:

44. [Not equal]: In September 2008, J. J. proved ...
45. [Not equal]: In October 2008, S. T. established ...
46. [Equal]: In November 2008, Z. A. proved ...

and so on. Woeginger provides links to research papers and, in some cases, further links to refutations of the claimed results.[12] The scorecard is evenly split, with twenty-five papers claiming $\mathcal{P} = \mathcal{NP}$ and twenty-four claiming $\mathcal{P} \neq \mathcal{NP}$. It makes for fascinating reading, but, thus far, none of the arguments has survived a serious review.

Nine of the twenty-five "Equal" results in Woeginger's list are established by delivering good algorithms for variants of the TSP. The methods employed vary from simple enumeration methods, to sophisticated attempts to obtain complete, but polynomial-sized, linear-programming formulations of the salesman problem. Each of the works appears to have serious flaws, but attacking the salesman is indeed an attractive route to proving $\mathcal{P} = \mathcal{NP}$.

Hamiltonian Circuits

If you are inclined to take up the search for a polynomial-time TSP algorithm, as part of a plan to earn $1,000,000, then it is useful to keep in mind that it is sufficient to focus on restricted versions of the problem. A common choice is to study methods for determining whether or not a graph has a Hamiltonian circuit. We know this variant of the TSP is

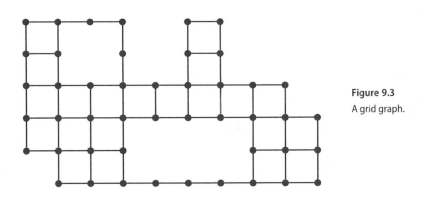

Figure 9.3
A grid graph.

itself $\mathcal{NP}$-complete. Even better, the Hamiltonian-circuit problem remains $\mathcal{NP}$-complete if we assume the input graph is bipartite, that is, the vertices of the graph can be colored either red or blue in such a way that each edge has a red end and a blue end. This specialization gives structure one can exploit in an algorithm, but, as far as polynomial solvability goes, it is just as difficult as the full TSP.

Alon Itai, Christos Papadimitriou, and Jayme Szwarcfiter pushed this further, proving that we can specialize the Hamiltonian-circuit problem to graphs that arise as finite subsets of the infinite grid of squares. An example of such a grid graph is displayed in figure 9.3, where we take a large rectangular grid and delete a subset of its vertices. Grid graphs present an attractive $\mathcal{NP}$-complete target for circuit-finding algorithms; David Johnson and Christos Papadimitriou call this the "ultimate special case of the TSP."[13]

Geometric Problems

The Hamiltonian-circuit problem is clean and tidy; no need to worry about evaluating travel costs. For most of us, however, it may be easier to develop intuition for the Euclidean version of the TSP, where the problem is specified by (x, y) locations and travel costs are equal to the straight-line distances between cities. Solving this version of the problem will also prove $\mathcal{P} = \mathcal{NP}$, and fetch \$1,000,000.[14]

Besides the big money, there is a second important open question in this setting: it is currently unknown if the Euclidean TSP is actually a member of the class $\mathcal{NP}$. This is at first sight shocking. The decision version of the problem asks if there is a tour of length at most a specified number K, so the natural certificate is a listing of the cities in an order that gives such a tour. We may assume each of the (x, y) coordinates are integers, so there is no problem representing the data. The technical difficulty arises from the square roots that must be computed. It is a simple matter to approximate any square root to as much precision as we need, but the sum of the square roots may add up to a value very, very close to K. It is not known if we can in polynomial time obtain a sufficiently accurate approximation to the tour length to decide if it is actually no larger than K.

Ronald Graham popularized this *sum-of-square-roots problem* in the early 1980s, and illustrated its possible difficulty with the following example. Add 1,000,000 to each of the numbers in the two lists below, then take the sum of their square roots.

$$1 \ 25 \ 31 \ 84 \ \ 87 \ \ 134 \ 158 \ 182 \ 198$$

$$2 \ 18 \ 42 \ 66 \ 113 \ 116 \ 169 \ 175 \ 199$$

For example, the first sum is $\sqrt{1000001} + \sqrt{1000025} + \ldots + \sqrt{1000198}$. These are innocent-looking computations, but the two resulting values

$$9000.449983568839730949026828861359029912$$
$$9000.449983568839730949026828861359029915$$

differ first in the 37th place after the decimal point. In general, it is not known if a polynomial number of digits is sufficient to decide if a sum of square roots is no larger than a supplied number K; here the size of the input is the sum of the number of digits in the (x, y) coordinates and the number of digits in K.

Settling the sum-of-square-roots problem is the direct route to proving the Euclidean TSP to be $\mathcal{NP}$, but it is not the only possibility. Perhaps there is an alternative means to show that a set of points has a tour of length at most K, making use of some, as yet unknown, geometric structure. This would be a very interesting development, and, as far as the TSP goes, perhaps more interesting than the important number-theoretic work needed in the sum-of-square-roots approach.

The Held-Karp Record

It will likely take a revolutionary idea to decide the polynomial solvability of the salesman, one way or the other. There is, however, the less lofty goal of slowly chipping away at the complexity of the problem by repeatedly improving the best-known running-time bounds for TSP algorithms. This gradual approach has the appealing aspect that methods achieving faster time bounds could be suitable in applications, enhancing the practical side of the TSP.

Faster and faster yet is indeed a good motto for complexity analysis, but for the TSP we seem to have reached a barrier with the 1962 work of Michael Held and Richard Karp.[15] This team's dynamic-programming algorithm solves any n-city TSP instance in time proportional to $n^2 2^n$, and this is where we still stand, after nearly fifty years. A revolution may be overstating what is needed to push beyond Held-Karp, but it clearly is going to take an exciting new idea.

Given the record status of Held and Karp, it is only fitting to give a full description of their algorithm. For the presentation, we consider an n-city problem, with cities named 1 through n and with travel costs denoted by $cost(1, 2)$, $cost(1, 3)$, and so on, for each pair of cities.

Fixing city 1 as the origin for the salesman, the Held-Karp solution is built from optimal subpaths, constructed for every subset of cities, excluding city 1, and for each possible ending point within the subset. As an example, take the subset $\{2, 3, 4, 5, 6\}$, with city 6 as the chosen ending point. A subpath for these cities is optimal, in the sense we mean here, if it is cheapest among all paths that begin at city 1, end at city 6, and visit cities 2, 3, 4, and 5, in any order, along the way. We denote the cost of such an optimal subpath as $trip(\{2, 3, 4, 5, 6\}, 6)$. To compute this value, we find the minimum of the four sums

$$trip(\{2, 3, 4, 5\}, 2) + cost(2,6)$$
$$trip(\{2, 3, 4, 5\}, 3) + cost(3,6)$$
$$trip(\{2, 3, 4, 5\}, 4) + cost(4,6)$$
$$trip(\{2, 3, 4, 5\}, 5) + cost(5,6)$$

corresponding to the possible choices for the next-to-last city in the subpath from 1 to 6, that is, we optimally travel to the next-to-last city then travel over to city 6.

This construction of a five-city $trip$-value from several four-city values is the heart of the Held-Karp method. The algorithm proceeds as follows. We first compute all one-city values: these are easy, for example, $trip(\{2\}, 2)$ is just $cost(1, 2)$. Next, we use the one-city values to compute all two-city values. Then we use the two-city values to compute all three-city values, and on up the line. When we finally get to the $(n-1)$-city values, we can read off the cost of an optimal tour: it is the minimum of the sums

$$trip(\{2,3,\dots, n\}, 2) + cost(2,1)$$
$$trip(\{2,3,\dots, n\}, 3) + cost(3,1)$$
$$\cdots$$
$$trip(\{2,3,\dots, n\}, n) + cost(n,1)$$

where the cost term accounts for the return trip back to city 1.

That is all there is to it. The running-time bound arises from the fact that in an n-city problem there are 2^{n-1} subsets that do not contain the origin. For each of these we consider at most n choices for the end city (actually, the number of choices is only the cardinality of the subset, but to make the counting easy we increase this to n), and the computation of the trip value involves fewer than n additions and n comparisons. Multiplying 2^{n-1} times n times $2n$, we see that the total number of steps is no more than $n^2 2^n$.

The running-time bound is better than checking all tours, once you get beyond ten cities, but it would be disappointing if Held-Karp is the best we can do. In looking to beat the record, one needs to focus on the 2^n term: replacing $n^2 2^n$ by $n2^n$ would not be considered an important step. But a bound of $n^2(1.99)^n$ or $n^2 2^{\sqrt{n}}$ would be newsworthy indeed, possibly signalling an era where future improvements could push us toward practical methods with strong running-time guarantees.[16]

Cutting Planes

If we want to beat Held-Karp, why not turn to the cutting-plane method? The writer/creator of xkcd.com, Randall Munroe, came right to this point in the publication of the TSP comic strip displayed in figure 1.5 in chapter 1. "What's the complexity class of the best linear programming cutting-plane techniques? I couldn't find it anywhere. Man, the Garfield guy doesn't have these problems."[17]

As the undisputed champion of practical computation, the cutting-plane method is indeed a natural candidate to consider in a complexity analysis of the TSP.

Unfortunately, it does not appear to be easy to get a handle on the worst-case performance of cutting planes. In 1987, Vašek Chvátal, Mark Hartmann, and I showed that a strong variant of branch-and-cut requires at least $\frac{2^{n/72}}{n^2}$ operations to solve a specially constructed nasty instance of the Hamiltonian-circuit problem; other TSP instances may require even more steps. The analysis, however, leaves open the possibility that new classes of cutting planes could result in a much better running-time bound. This is a nice theoretical target to complement the ongoing search for better-performing practical implementations of branch-and-cut.

Near-optimal Tours

A proof showing $\mathcal{P} \neq \mathcal{NP}$ would put to rest any hope for a good TSP algorithm, but it may still leave a window open for the salesman. For example, the tree-based heuristic of Nicos Christofides, described in chapter 4, is guaranteed to deliver a tour of cost no more than 1.5 times the cost of an optimal tour. What if this result could be improved to a 1.01-approximation algorithm, guaranteeing a tour within one percent of optimality? A fast computer implementation of such a method would be a fantastic practical tool in many applications of the problem.

Such approximation algorithms are definitely an interesting avenue to explore, even while P vs. $\mathcal{NP}$ remains undecided. To study these methods, however, we must restrict the allowed travel costs to exclude difficult yes/no problems.[18] The standard choice here, adopted by Christofides, is to assume costs are symmetric and satisfy the triangle inequality, that is, for any three cities A, B, and C, the cost to travel from A to B plus the cost to travel from B to C must not be less than the cost to travel from A directly to C. Problems meeting this natural condition are called *metric* instances of the TSP.

Christofides' algorithm first appeared in a Carnegie Mellon University research report in 1976. At the time, the result seemed easy enough. Thirty years later, with no improvements in sight, it no longer seems so simple. Indeed, it is a pressing open problem to find a polynomial-time α-approximation algorithm with α less than 1.5, capable of handing all metric instances.

To keep things fair and balanced, we must point out that it may, in fact, be impossible to defeat Christofides. On this negative side, Christos Papadimitriou and Santosh Vempala proved that, unless $P = \mathcal{NP}$, there can be no polynomial-time α-approximation algorithm for the metric TSP with α less than 1.0045.[19] Their work kills the idea of obtaining a super-good approximation method in the face of $P \neq \mathcal{NP}$, but it is not clear where the real computational barrier lies, closer to 1.0045 or closer to 1.5. It is disturbing that we cannot narrow this range, but you can add it to the list of interesting open research topics.

Arora's Theorem

Princeton's Sanjeev Arora proved a remarkable theorem that illustrates both the hopes and pitfalls of approximation methods. Arora showed that no matter what α we choose, as long as α is greater than 1.0, there exists a polynomial-time α-approximation algorithm for the Euclidean TSP.[20] Notice the contrast with the metric case, where, unless there exists a polynomial-time method to compute an optimal tour, we cannot hope for such a result. This is an interesting aspect of Arora's theorem, suggesting the Euclidean TSP may be more tractable than the general metric problem.

The potential pitfall is the following. Although Arora's theorem is a great theoretical result, the running times of the algorithms increase dramatically as α gets close to 1.0, and experimental results have been discouraging. This is a common feature of approximation methods, where very fine divisions of space are used to obtain high-quality solutions, at the expense of long search times. It remains an open problem to craft Arora's geometric technique into a practical TSP tool.

Do We Need Computers?

Neil Stephenson's science-fiction novel *Anathem* describes an exotic machine for solving the "Lazy Peregrin" problem, his fictional name for the TSP.

> "That's the one where a wandering fraa needs to visit several maths, scattered randomly around a map."
> "Yes, and the problem is to find the shortest route that will take him to all of the destinations."
> "I kind of see what you mean," I said. "One *could* draw up an exhaustive list of every possible route—"
> "But it takes forever to do it that way," Orolo said. "In a Saunt Grod's Machine, you could erect a sort of generalized model of the scenario, and configure the machine so that it would, in effect, examine all possible routes at the same time."

The Saunt Grod's machine is magic, but physical analogs of such all-tours-at-once devices have been conceived for, and in some cases tested on, the TSP. We should not forget that Turing-style computing is by no means the only tool for tackling the salesman problem.

DNA for the TSP

A biological candidate for Saunt Grod was proposed by University of Southern California professor Leonard Adleman in 1994.[21] Adleman is an award-winning computer scientist, well known as the "A" in the RSA cryptography system. His TSP device works at the molecular level, attempting to harness the immense information that can be stored in tiny amounts of DNA.

The variant of the TSP tackled by Adleman is a version of the Hamiltonian-path problem. Given as input a graph, the goal is to find a path to travel from a specified starting vertex to a specified finishing vertex,

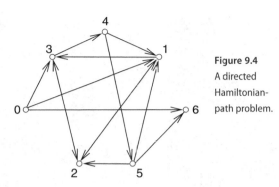

Figure 9.4
A directed
Hamiltonian-
path problem.

visiting every other vertex along the way; no travel costs are involved. The seven-city instance used in Adleman's DNA experiment is drawn in figure 9.4; the starting point is city 0 and the ending point is city 6. In this example, most of the edges model one-way streets and the Hamiltonian path through the vertices must obey the indicated directions; travel in either direction is permitted on the edges between vertices 1 and 2 and between vertices 2 and 3, while travel is permitted in only one direction on the remaining edges.

Adleman created a molecular encoding of the problem, assigning each of the seven vertices to a random string of twenty DNA letters. For example, vertex 2 was assigned to

$$TATCGGATCG|gtatatccga$$

and vertex 3 was assigned to

$$GCTATTCGAG|cttaaagcta.$$

By considering the twenty-letter tags as the combination of two ten-letter tags, indicated above by the groups of capital and lower case letters, an edge directed from vertex a to vertex b is represented by the string consisting of the second half of the tag for a and the first half of the tag for b. For example, an edge from vertex 2 to vertex 3 would be

$$gtatatccga|GCTATTCGAG,$$

taking the second group from vertex 2 and the first group from vertex 3. The only exceptions to this rule are edges involving the starting and ending points, 0 and 6, where the entire twenty-letter tag is used rather than just the first or second half. To capture cases where travel along either direction is possible, a pair of edges is created, one in each direction.

The next step in Adleman's approach is to provide a mechanism for joining edges together into a path. For each vertex, other than 0 and 6, the

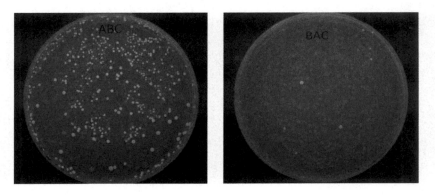

Figure 9.5
Bacterial computation of Hamiltonian paths. Images courtesy of Todd Eckdahl.

complementary DNA sequence for the tag is created. Such sequences act as "splints" to pair up edges in their proper orientation. For example, the complementary sequence for vertex 3 could serve to join the edges (2, 3) and (3, 4), since the first half of the sequence would pair up with the second half of the encoding of (2, 3) and the second half of the sequence would pair up with the first half of the encoding of (3, 4).

In the experiment, many copies of the DNA for edges and splints were produced in the laboratory and mixed together. After careful work over a seven-day period, a double strand yielding a Hamiltonian path was identified.

Bacteria

Adleman's experiment involved a week's effort in the laboratory as a mad scientist. An appealing alternative is to let a living organism handle the DNA manipulation, since that is their stock and trade. A research group of undergraduate students and faculty advisors from Davidson College, Johnson C. Smith University, Missouri Western State University, and North Carolina Central University did just this, setting up DNA in bacteria to solve a small instance of the Hamiltonian-path problem.[22]

Of course, working with bacterial computers presents its own challenge: how can one identify the DNA corresponding to a path if it is buried inside the living organism? The idea proposed by the research team is to use fluorescence properties to force a bacterial colony to light up if its DNA correctly identifies a path. They demonstrate the method on a three-city problem, where red and green fluorescence combine to produce yellow colonies when two directed edges line up in a Hamiltonian path.

The experimental results displayed in figure 9.5 show the idea. In the left-hand picture the DNA was initialized to be in the Hamiltonian-path order, and many yellow colonies are present after growth of the bacteria. In the right-hand picture the DNA was initialized in the wrong order, but after a sequence of mutations a number of colonies display the telltale yellow coloring.

Okay, three cities is not much of a TSP. The idea however is cool, and you have to start somewhere. The exponential growth of bacteria cells could possibly be utilized in larger computations, adopting more sophisticated methods for sorting through the produced colonies.

Amoeba Computing

Moving up the food chain, a research team in Japan demonstrated how a single-cell amoeba can be adopted to solve general instances of the TSP, not just Hamiltonian-path problems.[23] The centerpiece of their amoeba computer is displayed in figure 9.6. On the left is an amoeba and on the right a plastic structure with a star-like opening. An amoeba placed in the structure will, over time, modify its shape to fill the available star-like area. This shape-shifting process can be guided by turning lights on and off in each of the structure's radial arms, since an amoeba shrinks away from sources of light. The idea is to harness the amoeba's ability to optimize its shape in response to a changing environment.

An implementation of a four-city TSP solver on an amoeba computer is illustrated in figure 9.7. A star-like structure with sixteen radial arms is used, with each city represented by four arms. The arms for city A are labeled $A1$, $A2$, $A3$, and $A4$, indicating the position of city A in the tour; if the amoeba selects $A2$ then A is second in the tour order. An hour and fourteen minutes into the experiment, the amoeba begins to grow a branch in the A4 arm. At this point, the computer program turns on light sources

Figure 9.6
Unicellular amoeba and a barrier structure. Images courtesy of Masashi Aono.

7mm

7mm

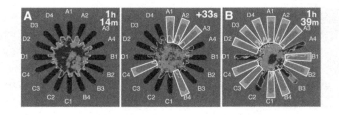

Figure 9.7
Amoeba solution
of a four-city TSP.
Images courtesy
of Masashi Aono.

at $A1$, $A2$, and $A3$, to enforce the rule that city A can only appear once in the tour. Also, lights are turned on at $B4$, $C4$, and $D4$, enforcing the rule that only one city can be fourth in the tour. These lighting steps are not enforced strictly during the shape-shifting process, giving the amoeba a chance to explore alternatives for the fourth position when lights are temporarily turned off. To guide the amoeba to short tours, after $A4$ is selected the cities furthest from A are periodically illuminated in positions 1 and 3, discouraging branches in these arms. Eventually the amoeba reaches a stable solution, corresponding to the tour D-C-B-A.

Again, four cities is not a legitimate computational challenge—the important part of the research is the creation of a new working computer with a biological component.

Optics

A light-based candidate for a Saunt Grod's machine made a big splash on the Web in the summer of 2007.[24] Applied to the Hamiltonian-path problem, the idea is to build a physical version of a graph, with fiber-optic cables as edges and delay devices as vertices. When the machine is in operation, light arrives at a vertex and is delayed for a fixed amount of time before being split into rays that are sent along each outgoing edge. The delay device serves as a signature to indicate that the light ray has passed through the individual vertex. Let D be equal to the sum of the delays incurred if we pass through each vertex exactly once. To solve the Hamiltonian-path problem, we send light through the starting vertex and check if light arrives, after D time units, at the finishing vertex.

This light-based machine appears to solve an n-city problem in a number of steps proportional to the size of the graph. This is what attracted all of the attention. A careful analysis by Romanian computer scientist Mihai Oltean, however, showed that in choosing the delays to give unique signatures, we necessarily have D at least as large as 2^n time units.[25] Current oscilloscopes cannot distinguish two rays arriving any closer than 10^{-12} seconds apart, so the time to solve a problem must be at least $2^n \times 10^{-12}$

seconds, an exponential growth rate. Still, the numbers are small enough that modest-size problems could be solved rather quickly.

Oltean estimates that a 33-vertex graph could be handled in one second. Not bad, but he also computes that implementing the delay devices in this case would require 8×10^{11} meters of cable. Moreover, the number of photons required for an instance of several hundred vertices would exceed the annual output of the sun. That begins to be problematic.

Quantum Computers

DNA, bacteria, amoeba, and optical TSP solvers all have an all-tours-at-once aspect, but they also require resources that grow exponentially with the number of cities. For a genuine Saunt Grod's machine, we may need to leave behind biology and classical physics. Indeed, a more likely candidate arises through the adoption of properties of quantum mechanics, first proposed for use in computing devices by Richard Feynman.

The basic component of a quantum computing device is the *qubit*, an unusual analog of the 0/1-bits used to represent information on classical computers. A qubit can hold the value 0 or the value 1, but it can also take on both these values simultaneously. Some magic via quantum mechanics gives the real possibility of examining all TSP tours at once: if we have 100 qubits, then together they can simultaneously encode 2^{100} possibilities.

Research groups around the world are working hard to solve the physics and engineering tasks that could lead to a functioning model of a quantum computer. Such a computer would give fast methods for factoring integers, via a famous result of Peter Shor, but it is a misconception that the salesman problem will be solved easily once we have access to a great many qubits. True, a million qubits would be enough to encode every tour through 1,000 cities, but there is a catch in the physics. Although all tours are represented simultaneously, when we actually examine the states of the qubits, all but one of the tours will vanish. Given this disappearing act, how can we guide the machine to select the best tour? It is not at all clear that this is possible.

Scott Aaronson, considering this point in a *Scientific American* article, describes possible limits of quantum computing.[26] He makes the following comment concerning the difficulty of producing polynomial-time quantum algorithms for $\mathcal{NP}$-complete problems.

> If such algorithms existed, however, they would have to exploit the problems' structure in ways that are unlike anything we have seen, in much the same way that efficient classical algorithms for the same

problems would have to. Quantum magic by itself is not going to do the job.

Quantum computers offer intriguing capabilities beyond Turing-style computing, but it remains to be seen if these can be harnessed to improve significantly our ability to route the salesman.

Closed Timelike Curves

Aaronson's article includes a discussion of several exotic computing models, including a favorite of mine, the time-traveling solver. The idea is simple enough. Start Concorde running on a reliable computer and set the program to deliver the solution, found sometime far into the future, back to the present time. Speedy, but a natural question is whether we can go ahead and turn off the computer once we receive the solution from our time machine. If we do, then how did Concorde find the solution we received?

People that ponder seriously these ideas consider a concept known as closed timelike curves, that is, paths through space and time that return to their starting point, forming a closed loop.[27] If such loops exist, then the salesman could be pursued along the curve, possibly allowing us to pick off the solution on the return journey.

Strings and Pegs

Coming back to earth, we should not fail to mention the physical device employed by Dantzig, Fulkerson, and Johnson, as well as by teams of assistants supporting actual salesmen in the early 1900s. Namely, a map with pins or pegs at the destination cities, and a string to lay out potential tours. This is a device to speed up by-hand computations, measuring the length of a tour by keeping a grip on the far end of the string. A bit easier to build than a time-traveling quantum computer, and the most practical physical TSP aid devised to date.

10: The Human Touch

Here is what we do. We take a group of talented young people, and we expose them to the history and theory of some famous $\mathcal{NP}$ problem. The traveling salesman problem will do nicely.
—Charles Sheffield, 1996.[1]

Salesmen, lawyers, preachers, authors, and tourists have been plotting tours for years, not to mention all of those tennis players collecting balls after long practice sessions. With all this experience, could the human mind be a viable non-computer platform for cracking the general TSP?

Humans versus Computers

Like any good sporting event, the 1997 chess match between World Champion Gary Kasparov and IBM's Deep Blue drew vocal supporters for both contestants. Those hoping to keep machines at bay for a few more years pulled for the human, while hardware and software fans aligned themselves with the computer. Science-fiction writer Charles Sheffield, covering the match for IBM, did not choose a favorite, but he was struck by the fact that a human could go toe-to-toe with a massive computing device on what is essentially a computational problem. He speculated whether humans could also compete successfully on other nasty computational challenges, such as the TSP.

Without specific training, as proposed by Sheffield, my money would be placed on the computer in a TSP showdown. One reason for this is a personal experience. At a mathematics workshop in 2007, Sylvia Boyd issued a fifty-city TSP challenge, with the rule that all calculations had to be by-hand only. The contest ran for a day and Dave Applegate and I produced the winning tour shown in figure 10.1, narrowly beating out fellow TSP researcher Gérard Cornuéjols. But, sadly, our tour was not optimal. After

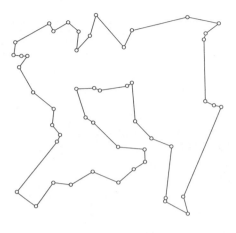

Figure 10.1
Winning tour from a
mathematics conference.

Figure 10.2
Smithsonian TSP exhibit. Photograph by Bärbel Klaaßen.

twenty years of research on the problem, we did not have the skill to match
what the computer could knock out in under a second.

Tour-finding Strategies

We may not have talent for finding optimal tours, but humans do quite well
in other aspects of the problem. The Smithsonian Air and Space Museum
has an exhibit that challenges visitors to find a tour through a selection of
airport locations in the United States. The display, shown in figure 10.2,
has a touch-screen computer monitor that allows a tour to be built step-
by-step, choosing one airport after another. In challenges such as this,

human subjects consistently produce good-quality tours for modest-sized examples. And although many mathematical methods can easily match these results, it seems clear that humans use a smaller number of explicit calculations in arriving at their good tours. This phenomenon has been studied by teams of psychologists, aiming to develop an understanding of human problem-solving abilities.[2]

Tour Gestalt

One of the findings of the psychology community is that high-quality geometric tours "feel right" when compared with tours of lesser quality, perhaps indicative of human desire for minimal structures. This theme was emphasized in an experiment led by Douglas Vickers at the University of Adelaide in Australia.[3] In the study, two groups were presented with identical 10-city, 25-city, and 40-city instances of the TSP (two of each size), but with different instructions on how to proceed. The *Optimization* group was asked to find the shortest tour in each example, while the *Gestalt* group was asked to find a tour such that "the overall pathway looked most natural, attractive, or aesthetically pleasing." The results showed a striking similarity in the quality of tours obtained by the two groups. The champion tour finder in the test was in fact a fashion designer belonging to the Gestalt group; she produced the shortest tours in five of the six examples.

Tours Found by Children

A study led by Iris van Rooij at the University of Victoria in Canada examined how children perform in tour-finding experiments, compared with adults facing the same examples.[4] This approach gave researchers a means to consider perceptual versus cognitive skills, since young children would be expected to rely primarily on their perception of good structure when searching for tours.

The participants in the study were 7-year-old and 12-year-old elementary school students, and a group of university students; the elementary school participants received a sticker as a reward for their work. The TSP test set consisted of randomly generated 5-city, 10-city, and 15-city instances, including five of each size; the performance was measured as the percentage by which the lengths of the tours produced exceeded optimal values. The results, displayed in table 10.1, indicate an improvement in performance as we move from children to adults, but even young children obtained reasonably short tours.

Table 10.1
Average percent over optimal tour lengths found by children and adults.

Number of Cities	7-Year-Olds	12-Year-Olds	Adult
5	3.8%	2.5%	1.7%
10	5.8%	3.4%	1.7%
15	9.4%	5.0%	2.7%

The Convex-hull Hypothesis

James MacGregor and Thomas Ormerod at Lancaster University in England focused on the degree to which the global shape of a set of points serves as a guide in tour finding.[5] A measure of this shape can be obtained by considering how a rubber band will enclose the set of cities, as in figure 10.3. The curve traced by the rubber band is the border of the convex hull of the cities. The border is usually not itself a tour, but it is easy to check that, in order to avoid crossings, an optimal tour must trace the border cities in the order in which they appear as we walk around the border. This convex-hull rule is indicated in figure 10.4 with the optimal tour for the 33-city Procter & Gamble TSP: the twelve points on the border appear in the same order as in the optimal tour, and the remaining cities are picked up by moving in from the border on short subpaths. In this context, cities not on the border are called interior points.

Following a detailed analysis of experimental results on human solutions to 10-city and 20-city examples, MacGregor and Ormerod conclude

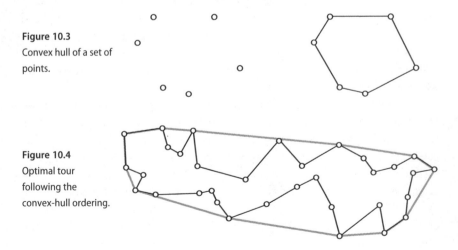

Figure 10.3
Convex hull of a set of points.

Figure 10.4
Optimal tour following the convex-hull ordering.

Figure 10.5
A human TSP
experiment.
Photograph courtesy
of Jan Wiener.

that the complexity of finding an approximate solution to an instance of TSP is determined by the number of its interior points. Moreover, the researchers write the following: "The evidence presented here indicates that human subjects reach solutions based on the perception of global spatial properties of a TSP, and in particular, of the boundary of the convex hull." The degree to which this hypothesis holds is the subject of a lively debate in the human-problem-solving community. Much of the technical discussion centers on the types of data sets used in experimental studies, but there is also the general question of whether humans make use of global-to-local strategies or rather local-to-global strategies. In the former we perceive an overall structure and then make local choices to fit the cities into this structure, while in the latter we carry out local analysis (such as clustering points) and then try to best put the local information into an overall tour.

Physical TSP Instances

The TSP instances in these human studies are visual, in the sense that the problem is presented as tracing a tour on a piece of paper or on a computer screen. Jan Wiener and a team at the University of Tübingen in Germany contrasted this with human performance on a physical TSP, where participants were asked to navigate a 6.0 × 8.4 meter room to visit a list of targets.[6] In their experiment, twenty-five pillars were arranged with a colored symbol on each pillar. The participants were given a start location and a list of up to nine symbols that they should visit before returning to the start. The results again support the claim that humans are quite good at solving small instances of the TSP—in this case for a wage of eight euros per hour.

The TSP in Neuroscience

Examples of the TSP that are either very small or constructed with a large proportion of the cities on the convex-hull border are routinely solved by humans, with little variation among study participants. Individual differences in performance arise quickly, however, when general problem instances have twenty or more cities, as the fashion designer in the study of Vickers et al. demonstrates. On even larger examples, having fifty cities, a second study led by Vickers found consistent differences in the tour quality produced by individuals.[7] The researchers also noted a modest correlation between TSP ability and scores on a standard nonverbal intelligence test.

Trail Making

Differences in human performance on TSP-like problems have long been a resource for clinical studies in neuropsychology. A prime example is the Trail Making test from the Halstead-Reitan Battery.[8]

The first part of Trail Making consists of the twenty-five labeled cities displayed in the left-hand side of figure 10.6. The test is administered by asking the subject to draw a path connecting the cities in consecutive order, requesting that the drawing be completed as quickly as possible and pointing out errors as they occur. The correct path, displayed in the right-hand side of figure 10.6, is clearly not an optimal route through

Figure 10.6
Trail Making (Part A).

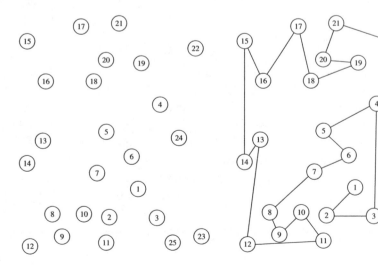

the cities, but it is a relatively short route without crossings. A second part of Trail Making consists of a similar task, but where the cities are labeled $1, A, 2, B, \ldots, 12, L, 13$. The two-part test was developed by US Army psychologists in the 1940s; the commonly used scoring system introduced by Reitan is based entirely on the times taken by the subject to complete the tasks.

Numerous experiments point to the sensitivity of Trail Making in identifying patients suffering from brain damage. In fact, a 1990 survey singled out Trail Making as the most widely used test among members of the International Neuropsychological Society.[9] It is interesting that Trail Making always makes use of the specific distributions of city locations given in the two parts of the test; there does not appear to be a reliable method to generate alternative placements of the cities with good clinical properties.

Animals Solving the TSP

Humans are pretty good TSP solvers, so how about beasts? The question was posed in the 1973 study by Emil Menzel, working with a team of chimpanzees.[10] Menzel devised an ingenious scheme to coax the subjects into traveling along an efficient tour, since neither a payment of eight euros nor a reward of a sticker would likely do the job. To begin a trial, six chimpanzees were kept in a cage at the edge of a field. A trainer would take a selected animal from the cage and carry it around the field while an assistant hid eighteen pieces of fruit at random locations. The animal was returned to the cage and after a two-minute waiting period all six chimpanzees were released. The selected animal would make use of its memory of the food locations to quickly gather the treats before the other animals located them by simple foraging.

The route taken by Bido, one of the chimpanzees in the study, is depicted in figure 10.7. Bido began at the point marked "Start" on the boundary of the field and finished at the location marked "Finish"; the arrows on some of the links indicate the direction of travel. The chimpanzee missed four pieces of fruit, but altogether Bido made a remarkably good tour working from its memory of the food locations.

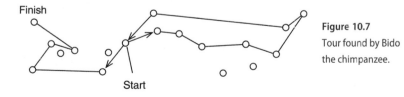

Figure 10.7
Tour found by Bido the chimpanzee.

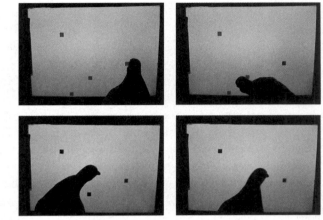

Figure 10.8
Pigeon solving a TSP.
Images courtesy of
Brett Gibson.

Other studies of animals solving the TSP include vervet monkeys, marmosets, and rats. The reported experiments all involve physical instances of the problem, with animals gathering food from scattered locations. A different approach was taken in a pigeon study led by Brett Gibson of the University of New Hampshire.[11] In his work, birds were trained to select Hamiltonian paths by pecking locations displayed on a touch screen monitor, as shown in figure 10.8. When all displayed cities were visited, the pigeon received two food pellets. The birds selected significantly shorter paths than random choices would produce, but typically longer than a nearest-neighbor approach. To encourage the birds to find shorter tours, a second experiment was carried out where food pellets were given only if the constructed solution was of sufficiently high quality. Not to be denied their snacks, the pigeons stepped up and developed good strategies for the small TSP instances in the test set.

In each of these studies, the simplicity of the rules for the problem allowed researchers to set up non-trivial experiments, and the TSP task provided a good means for testing the spatial cognitive skill of the various animal participants.

11: Aesthetics

*Mathematics was always a source of complex
(for outsiders, mystical) patterns.*
—Jaroslav Nešetřil, 1993.[1]

When a mathematician refers to a particular item of study as beautiful, it comes without any implication that the beauty can be realized in a physical form. This holds true for the TSP. A tour through a set of points may have a pleasing shape, but it is the combined beauty of the geometry and complexity of the problem, not the tour itself, that attracts mathematicians. Nevertheless, the TSP has been adopted in several engaging works of art, in some cases successfully capturing the mathematical essence that has brought so much attention to the salesman.

Julian Lethbridge

I was delighted to discover the *Traveling Salesman* paintings by Julian Lethbridge. Lethbridge is a celebrated contemporary artist, whose work is included in collections of the National Gallery of Art in Washington, the Metropolitan Museum of Art in New York, the Art Institute of Chicago, and the Tate Gallery in London. An impression of his style can be gathered from the following comments.

> Julian Lethbridge seems determined to demonstrate all the things you could possibly do with intersecting curves.
>
> —*New York Times*, 1995.[2]

> Lethbridge's abstraction is cerebral, often based on mathematical or natural principles.
>
> —*ULAE*, 1997.[3]

Figure 11.1
Traveling Salesman,
Julian Lethbridge, 1995,
lithograph, 43.75 × 42
inches. Image courtesy
of Julian Lethbridge
and United Limited Art
Editions.

His vocabulary often originates from random or naturally occurring
patterns, such as shattered glass or a spider's web.

—*Paula Cooper Gallery*, 1999.[4]

Straight, curved, gridded, circular, organic, ordered, chaotic, in-
cised, built up, drawn, painted, assertive, delicate—these kinds of
lines and more are the basis of Lethbridge's pictorial systems, and
are the elements with which he constructs his portrayals of nature.

—*Art in America*, 2007.[5]

The remarks hint at concepts familiar to mathematicians. Indeed, in a
meeting with Mr. Lethbridge in New York, our discussion began on the
topic of his artwork, but moved quickly to general mathematics, where
Lethbridge has interest in, and intuition for, the aesthetics of our discipline.
When discussing his *Traveling Salesman* series, Lethbridge explained
that he came upon a description of the TSP in a journal and was struck
immediately by the economy of thought and space provided by good
solutions to the problem. Two beautiful examples from his series are
displayed in figures 11.1 and 11.2. Reviewers in the *New York Times* and
the *Baltimore Sun* compare these to the map paintings of Jasper Johns
and to Jean Dubuffet's *Compagnie Fallacieuse*.[6] Such references would be

Figure 11.2
Traveling Salesman 4,
Julian Lethbridge,
1995, oil on linen, 72 ×
72 inches. The Robert
and Jane Meyerhoff
Collection, photograph
by Adam Reich.

disappointing to fans of the TSP, however, since neither of the cited works displays the organization of space provided by the tour in Lethbridge's images.

In the first of the two paintings, note that the non-crossing tour forms a simple closed curve, known as a Jordan curve, partitioning space into two regions, one bounded and the other unbounded; the bounded region is the one we think of as the interior of the curve. The painting captures this partition, using textures to highlight the interior and exterior, separated by a white rendering of the tour.

The same tour is displayed in a much different way in the second piece. Here the myriad of choices presented in alternative paths through the point set is suggested by multiple paint strokes radiating from unseen city locations. This large painting is part of the Meyerhoff Collection of the National Gallery of Art.

Jordan Curves

A tour drawn on a sphere would not have an interior or exterior, just two regions, linked together like pieces in a puzzle. This symmetry between the sides of a Jordan curve is apparent in Robert Bosch's *Embrace* sculpture, shown in the photograph displayed in the left-hand side of figure 11.3.

Figure 11.3
Left: *Embrace* TSP sculpture. Right: Symmetric rings. Images courtesy of Robert Bosch.

This work is not on a sphere, and the outer ring makes it clear which of the two regions is actually the exterior, but the symmetric view from the center is striking. *Embrace* was honored by the American Mathematical Society and the Mathematics Association of America in 2010, receiving the Mathematical Art Exhibition First Prize.

The artist is the same Robert Bosch who created the 100,000-city Mona Lisa TSP described in chapter 1. Bosch is a research mathematician and the current chairman of the Department of Mathematics at Oberlin College. His Ph.D. thesis is in mathematical optimization, and Bosch wields this as a tool in his art.[7]

> After I get an idea for a piece, I translate the idea into a mathematical optimization problem. I then solve the problem, render the solution, and see if I'm pleased with the result. If I am, I stop. If not, I revise the mathematical optimization problem, solve it, render its solution, and examine it. Often, I need to go through many iterations to end up with a piece that pleases me. I do this out of a love of mathematical optimization—the theory, the algorithms, the numerous applications.

One of the optimization methods favored by Bosch is the use of exact and heuristic algorithms for the TSP. The Mona Lisa image from chapter 1 and a drawing of Botticelli's *The Birth of Venus*, presented in the next section of this chapter, are two of the numerous *TSP Art* pieces created by Bosch with his optimization tools. Together with University of Waterloo computer scientist Craig Kaplan, Bosch has developed sophisticated techniques for placing cities in such a way that a good tour produces an interesting rendering of an original image.[8]

The Bosch-Kaplan process will create a working TSP drawing of the interlocking rings that make up the *Embrace* sculpture, but the resulting Jordan curve may put adjacent arms of the rings into the same region of space, either the interior or the exterior. This is where Bosch steps in with a problem revision, employing an integer-programming technique he calls "bending the curve to our will."[9] Bosch selects two points in space that he would like to have on opposite sides of the Jordan curve formed by the TSP tour. This geometric requirement is equivalent to stating that the line segment joining the two points must cross the TSP tour an odd number of times, and this is a condition that can be modeled as an additional constraint in an integer-programming formulation of the TSP. So Bosch adds the side constraint, solves the new optimization problem with an integer-programming code, and takes another look at the resulting Jordan curve.

The *Embrace* sculpture is created from quarter-inch-thick pieces of metal; the inside region is stainless steel and the outer region is brass. A water-jet cutter was used to cut along the 726-city TSP tour in each of the two types of metal, creating two versions of *Embrace* by pairing the copies of the two regions. In the cutting process, a narrow band of metal was removed from each of the pieces, resulting in the empty space that highlights the path through the sculpture.

Bosch also produced a larger version of a symmetric-ring Jordan curve, displayed in the right-hand side of figure 11.3. Concerning this piece, Bosch writes the following in an e-mail message.

> I added side constraints to the TSP to force the edges of the tour to have 5-fold rotational symmetry in both the center of the circle and near its edge, and 10-fold rotational symmetry in between. I added additional side constraints to achieve the appearance of interlacing when the inside and outside were colored.

The TSP in this case contains 2,840 cities, making it difficult to solve exactly the resulting integer-programming problems; the tour used in the image was created with a heuristic algorithm built into the software.

Philip Galanter's TSP Murals

Bosch and Lethbridge each make subtle use of the TSP and the resulting Jordan curves: in Bosch's case the subtlety comes from symmetry in design and in Lethbridge's case the subtlety arises in texturing of the painting. A more stark approach is employed by Philip Galanter in a series of large TSP murals. An example of Galanter's work is displayed in the photograph

Figure 11.4
TSP mural. Courtesy
of Philip Galanter.

in figure 11.4, where a TSP tour is used to paint a wall in two highly contrasting colors.

Galanter is an assistant professor in the Department of Visualization at Texas A&M University; his work centers around the use of complexity and automation in art and music. Concerning his TSP pieces, Galanter made the following comments in an introduction to his work at an exhibition in Lima, Peru.[10]

> The Traveling Salesman mural series explores emergence in a different way. The design for each mural is uniquely generated for each specific wall. A large number of random points are generated, and then a computer program calculates the optimally shortest line that connects each point once and only once. Somewhat mysteriously, each resulting mural exhibits a shared visual style. In nature the beauty of a tree or a plant is the result of nature solving an optimization problem; how can the plant gather as much sunlight as possible using the minimum amount of organic resources? Both optimization examples suggest that beauty is not meaningless, nor is it the exclusive province of individual genius. Beauty is posited as being both meaningful and publicly understandable.

Figure 11.5
incredulous MJ.
Copyright J. Eric Morales,
www.labyrinthineprojection.com.

Galanter plans for a number of further pieces in his TSP series, including a 16-foot-diameter metal sculpture and a tour drawn on a large campus lawn using equipment designed for marking lines on football fields.

Continuous Lines

Portland-based artist J. Eric Morales, known professionally as *Mo*, uses hand-drawn Jordan curves in an art form he calls Labyrinthine Projection. Drawing with one continuous line, Mo is able to capture intense expressions in his portraits. His technique has received considerable attention, with pieces appearing on popular Nike shoes and iPod covers, and in drawings commissioned by star basketball player Michael Jordan.

Mo writes that as a boy he would spend endless hours filling the screen of an Etch-A-Sketch with a nonintersecting line that meandered randomly over the toy's surface. The idea returned to him while studying art and learning of the formal principles of line density and light value; he realized that he could create images of photo-like quality by modulating a continuous line's distance from itself—closer for dark areas and further for lighter ones. The resulting curves have clear similarities to short TSP tours.

Mo has many ideas for extending his TSP-like work into other media. His latest work uses sunlight passing through labyrinth-inspired sculptural objects that project shadow images of recognizable faces. In another application he is developing Labyrinth Man, a headless humanoid with

Figure 11.6
Botticelli's *The Birth of Venus* as a TSP. Courtesy of Robert Bosch.

translucent skin animated by a morphing color labyrinth that visually mimics its environment and represents text on its surface to communicate.

Bosch-Kaplan Drawings

In an e-mail letter, Morales comments on a connection with the Bosch-Kaplan TSP Art mentioned in the previous section.

> I am quite familiar with the TSP, having been acquainted with Craig Kaplan many years ago who came across a $70' \times 40'$ Labyrinthine Projection of professional skater Paul Rodriguez I created for Nike in Los Angeles for the X-Games. Kaplan's algorithm is particularly notable to me as it is the computer generated solution that most resembles my hand-drawn process.

The TSP Art project aims to create continuous-line reproductions of original images, using a heuristic algorithm to produce short tours through carefully selected sets of city locations. A 140,000-city example is displayed in figure 11.6.

The project began at Oberlin College, where Bosch and student Adrianne Herman developed a procedure for placing cities in proportion to the various shades of gray in a digital image. More precisely, Bosch and Herman divide an image into a grid and place between 0 and k cities at random locations in each grid cell, depending on the average level of gray in the cell, where 0 is nearly white and k is nearly black; the value of k and the size of the grid control the number of cities in the resulting TSP.

A short tour through a Bosch-Herman point set produces a drawing where the original image can be recognized, but the clustered points often create jagged paths that do not capture well the continuous color tone of the original. Bosch and Kaplan greatly improved this aspect of the drawings, adopting a multiple-round algorithm for distributing the cities, based on computing centers of gravity. In their calculations, the tone of the original image is used to weight the geometric distances, so that points migrate to dark regions. Overall, the process avoids abrupt transitions in the density of the cities, allowing for subtle transitions from dark to light tones.[11]

Art and Mathematics

The epigraph of this chapter is taken from an essay by Jaroslav Nešetřil exploring connections between the thought processes of artists and mathematicians. Nešetřil, from Charles University in Prague, is a leading researcher in the field of discrete mathematics, as well as a successful artist, together with his longtime collaborator, the professional and well-known artist Jiří Načeradský. An example of Načeradský and Nešetřil's work is displayed in figure 11.7; this piece features a continuous curve wandering in three-dimensional space.

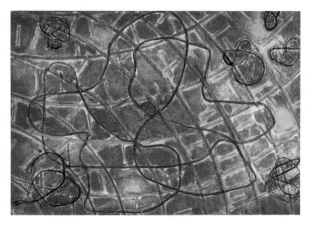

Figure 11.7
Jiří Načeradský and Jaroslav Nešetřil, 1997, mixed media, 100 × 120 cm. Image courtesy of Jaroslav Nešetřil.

Figure 11.8
Research Institute for
Discrete Mathematics
and Arithmeum, Bonn,
Germany. Image courtesy
of Bernhard Korte.

In his essay on connections, Nešetřil notes the unifying idea expressed by the Church-Turing Thesis in mathematics, which states that all algorithmic work takes a common form, captured by the execution of a Turing machine. He speculates whether a similar *Creative Thesis* may hold for general human activity. "All sufficiently deep activities, all sufficiently deep understandings have profound similarities. This is exhibited in the way the work (knowledge) is organized, in the way it is revealed and in the way it interacts with other activities".[12] Nešetřil draws his thesis from parallel developments in art and mathematics over the past two centuries, where both fields freed themselves of long-existing constraints, only to see new common forms arise, such as surrealism, on the art side, and modern set theory, on the mathematics side.

Constructivist Art and VLSI

Nešetřil has been a long-term visitor to the Research Institute for Discrete Mathematics in Bonn, Germany. The director of the institute, Bernhard Korte, is another person who lives in both the mathematics and art worlds. The Bonn institute is a top center for discrete optimization, but it also houses the *Arithmeum* museum for computing, art, and music. A photograph of the institute building is displayed in figure 11.8; the layout of the floors has mathematics researchers working among the museum's collections.

The Arithmeum includes numerous pieces of particular interest to mathematicians. Indeed, Bernhard Korte explains a museum focus on constructivist art as follows. "First and foremost we must confess that we

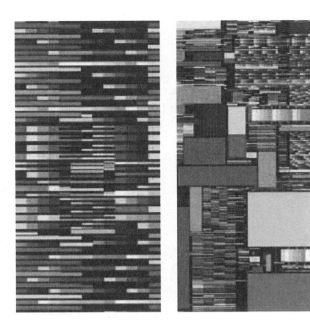

Figure 11.9
Design of the *Philipp*
VLSI chip. Images by
Ina Prinz.

feel an affinity for geometric and constructivist art forms. Why? Perhaps because absolute geometric forms combined with colours chosen from a basic colour scale are balsam to the naive soul of a mathematician".[13] The Bonn collection includes work of Josef Albers, Max Bill, Jean Gorin, Richard Paul Lohse, Leon Polk Smith, and Charmion von Wiegand, to name just a few. It is a great pleasure to visit the institute and study while surrounded by the beautiful exhibits; the unique setting is a clear contributor to the academic vitality of discrete mathematics in Bonn.

In the realm of applications, Bonn's speciality is the optimal design of integrated circuits, that is, the design of computer chips that make up the heart of modern electronic devices. This area is known as very-large-scale integration (VLSI) and Bonn researchers are the world's leaders in applying discrete mathematics to improve the speed and organization of the billion or so transistors that make up complex VLSI chips. The engineering involved in this work is overwhelmingly complex, but the end results can be aesthetically pleasing, without any knowledge of the mathematics behind the scenes. Arithmeum catalogs make this point clear with numerous interesting images obtained from completed VLSI projects. Two examples of this work are displayed in figure 11.9; the geometric patterns are representations of the layout of components in a particular computer chip, while the range of colors was selected by museum director Dr. Ina Prinz.

Nešetřil makes the following observation on the VLSI design process, bringing this applied research into the framework of his Creative Thesis.[14]

Chip design represents one of the most concentrated of human activities. This activity is interdisciplinary with methods spanning computer science, mathematics, physics and even philosophy. It is no wonder that the interplay of these activities displays some affinity to artistic works.

We can express similar feelings for the traveling salesman problem, and hope to see further connections between the TSP and art, as progress is made in understanding the problem's fundamental complexity.

12: Pushing the Limits

*The problem is certainly not a closed one, and I hope more research
will be done, both toward finding better computational methods,
and toward a better mathematical understanding of the problem.*
—Delbert Ray Fulkerson, 1956.[1]

T he beauty of the TSP will no doubt continue to attract mathematicians
and computer scientists for years to come.

> Christos Papadimitriou told me that the traveling salesman problem
> is not a problem, it's an addiction.
>
> —Jon Bentley, 1991.[2]

> It's addictive. No matter how much progress you make, you always
> have the nagging feeling that you still did not nail down a couple of
> hunches that could bring about another quantum leap.
>
> —Vašek Chvátal, 1998.[3]

We offer no tips for breaking a TSP addiction. Far from it. I wouldn't
hesitate to include small TSP challenges on the backs of candy wrappers,
if given a chance.

Although there is no candy on the menu, throughout the book we have
touched upon open research questions, such as the Mona Lisa and World
TSP challenges, the 4/3rds Conjecture, beating the Held-Karp running-
time bound, and improving Christofides' approximation barrier. Bouncing
around ideas for such topics is fun, but I don't want to disguise the fact
that long hours may be needed for a breakthrough: fuller understanding of
the TSP will only come through a passionate desire to dig deeply into the
computational mystery surrounding the problem.

The Role of the Salesman

Computer-science giant Avi Widgerson has suggested links between complexity theory and potential limits of human knowledge. Indeed, if it is shown that $\mathcal{P} = \mathcal{NP}$, then a new era will be ushered in, with efficient computational tools to model and understand the world around us. On the other hand, if $\mathcal{P} \neq \mathcal{NP}$, as most experts expect, then numerous important questions may go unanswered indefinitely: increasingly fast computing machines can never overcome an exponential rate of growth in the running time of a solution method.

So how should we address the challenges ahead? One answer lies in the take-no-prisoners approach adopted in computational studies of the TSP. If $\mathcal{P} \neq \mathcal{NP}$, then there are limits to general-purpose solution techniques, in science and elsewhere. But what are these limits and how widely do they constrain our quest for knowledge? The salesman can play a crucial role in this context, demonstrating whether or not focused efforts on a single, possibly unsolvable, problem will produce results beyond our expectations.

Let's close on this note, with the hope that readers may be encouraged to take up the study of the TSP, both with an eye on the million-dollar complexity prize and on the practical, step-by-step, computational attack. The traveling salesman problem is as tough as it gets, but, as Rashers Ronald would say, bash on regardless.

Notes

Chapter 1. Challenges

1. IBM Corporation press release from January 2, 1964, describing a new computer program for solving small instances of the traveling salesman problem. The computer program was developed by Michael Held, Richard Karp, and Richard Shareshian.
2. Menger (1931).
3. TOP500 Supercomputer List, June 2009.
4. Little, J. D. C., et al. 1963. Operations Research **11**, 972–89.
5. *Newsweek*, July 26, 1954, page 74.
6. Karg, R. L., G. L. Thompson. 1964. Management Science **10**, 225–48.
7. Charles Stross's story "Antibodies" appears in the collection *The Year's Best Science Fiction* edited by G. Dozois, St. Martin's Press, New York, 2001.
8. Flood, M. M. 1956. Operations Research **4**, 61–75.
9. Edmonds, J. 1967. J. Res. Nat. Bur. Stand. Sec. B **71**, 233–40.
10. Karp (1972).
11. One can argue that we must also consider the precision of the travel costs: if we need to read millions of digits to obtain the cost of travel between city x and city y, then this must also be considered in measuring the size of the problem. We can, however, safely skip over this detail in our analysis. Indeed, the salesman problem with each cost an integer no larger than some fixed constant K is plenty difficult and it captures the general complexity of the problem.
12. Fortnow, L. 2009. Communications of the ACM **78**, 78–86.
13. The term *algorithm engineering* goes back at least to 1997, when the first Workshop on Algorithm Engineering was held in Venice, Italy. A German Science Foundation (DFG) research program devoted to the topic describes the field as consisting of the "design, analysis, implementation and experimental evaluation of practical algorithms." One of the program leaders, Petra Mutzel, holds the Chair of Algorithm Engineering at the Technical University of Dortmund.
14. The T-shirt in the photograph was worn by Jessie Brainerd's fellow student Bill Kay at a Halloween party in Budapest. The blog entry for the party mentions two students dressed as P vs. $\mathcal{NP}$, armed with dart guns to battle one another.

15. Rashers Ronald appears in J. P. Donleavy's book *The Destinies of Darcy Dancer, Gentleman*, Atlantic Monthly Press, 1994.

Chapter 2. Origins of the Problem

1. Dantzig, Fulkerson, and Johnson (1954).
2. The list of towns in Maine is part of a large collection of bills and letters from H. W. Cleveland to the Page Seed Company. I was lucky to be able to purchase the collection on eBay. Not all of my buys turned out to be so interesting—I also purchased 50 years of annual diaries written by a salesman, only to learn that his tours consisted of trips through five or six cities around Syracuse, New York.
3. *Der Handlungsreisende—wie er sein soll und was er zu thun hat, um Aufträge zu erhalten und eines glücklichen Erfolgs in seinen Geschäften gewiss zu sein— Von einem alten Commis-Voyageur.* B. Fr. Voigt, Ilmenau, 1832.
4. Translated from the German original by Linda Cook.
5. These are the first lines of the poem "The Drummer" from the book by Marshall, G. L. 1892. *O'er Rail and Cross-ties with Gripsack. A Compilation on the Commercial Traveler.* New York, G. W. Dillingham.
6. Fraker, G. C. 2004. Journal of the Abraham Lincoln Association **25**, 76–97.
7. Hampson, J. 1791. *Memoirs of the late Rev. John Wesley.* J. Johnson, London, UK.
8. Banks, N. 1830. *The Life of the Rev. Freeborn Garrettson: Compiled from his Printed and Manuscript Journals, and other Authentic Documents.* New York. Published by J. Emory and R. Waugh.
9. Hibbard, B. 1825. *Memoirs of the Life and Travels of B. Hibbard: Minister of the Gospel, Containing an Account of his Experience of Religion.* New York. Published by the author.
10. Gribkovskaia, I., O. Halskau, G. Laport. 2007. Networks **49**, 199–203.
11. Euler, E. 1741. Comm. academiae scientiarum Petropolitanae **8**, 128–40.
12. An English translation of Euler's paper, together with a detailed scholarly presentation of its influence, can be found in the book by Biggs et al. (1976).
13. Euler, L. 1766. Mémoires de l'Academie Royale des Sciences et Belles Lettres, Année 1759, Berlin. 310–37.
14. Hamilton, W. R. 1856. In: H. Halberstam, R. E. Ingram, eds. *The Mathematical Papers of Sir William Rowan Hamilton*, Volume III. Cambridge University Press, Cambridge, UK. 612–24.
15. The i operation is to reverse the travel along an edge; κ is to rotate counterclockwise to the next edge at a vertex; and λ is to turn left at a vertex. So, for example, applying λ five times will trace a pentagon in the graph: since this brings us back to our starting point, Hamilton writes $\lambda^5 = 1$.
16. The two images of Hamilton's games are courtesy of James Dalgety, whose Puzzle Museum contains a fascinating collection of games and puzzles, both old and new.

17. This letter from Hamilton to De Morgan, dated October 26, 1852, is reported in an excellent biography of Hamilton by Thomas L. Hankins: *Sir William Rowan Hamilton*, Johns Hopkins University Press, 1980.
18. Menger (1931).
19. Dantzig, Fulkerson, Johnson (1954).
20. Flood, M. M. 1956. Operations Research **4**, 61–75.
21. Flood, M. M. 1984. The Princeton Mathematics Community in the 1930s, Transcript Number 11 (PMC11). Princeton University.
22. Many of the details of Menger's work are reported in Schrijver (2003).
23. Hoffman and Wolfe (1985).
24. Reid (1996).
25. Reid (1996).
26. Mahalanobis (1940).
27. Ghosh, M. N. 1949. Calcutta Stat. Assoc. **2**, 83–87.
28. The result states that with probability 1, as n gets very large, the optimal tour length divided by $\sqrt{n}$ will approach β. Beardwood, J., J. H. Halton, J. M. Hammersley. 1959. P. Camb. Philos. Soc. **55**, 299–327.
29. An interesting study of β by physics researchers is given in Percus, A. G., O. C. Martin. 1996. Phys. Rev. Let. **76**, 1188–91.

Chapter 3. The Salesman in Action

1. Albers and Reid (1986).
2. Bartholdi III, J. J., et al. 1983. Interfaces **13**, No. 3, 1–8.
3. Suri, M. 2001. SIAM News **34**, p. 1.
4. Dosher, M. 1998. *Wall Street Journal*, February 9, B1.
5. Agarwala, R., et al. 2000. Genome Research **10**, 350–64.
6. Carlson, S. 1997. Sci. Am. **276**, 121–24.
7. Kolemen, E., N. J. Kasdin. 2007. Adv. Astronaut. Sci. **128**, 215–33.
8. Bland, R. G., D. F. Shallcross. 1989. Op. Res. Let. **8**, 125–28.
9. Grötschel, M., M. Jünger, G. Reinelt. 1991. Zeit. Op. Res. **35**, 61–84.
10. Lenstra, J. K. 1974. Operations Research **22**, 413–14.
11. Climer, S., W. Zhang. 2006. J. Mach. Learn. Res. **7**, 919–43.
12. References can be found in Applegate et al. (2006), Section 2.7.

Chapter 4. Searching for a Tour

1. Karg, R. L., G. L. Thompson. 1964. Management Science **10**, 225–48.
2. Dantzig et al. (1954).
3. Throughout the book we focus on the symmetric form of the TSP, where costs do not depend on the direction of travel, that is, the cost to travel from city A to city B is the same as to travel from B back to A. This restriction is mainly an attempt to control the length of our discussions, but please do not feel that

you are being cheated: it is an exercise to show that any TSP instance with n cities can be converted to a $2n$-city TSP with symmetric costs.

4. Dantzig et al. (1954).
5. Throughout the book log values are base 2.
6. Robacker, J. T. 1955. RAND Research Memorandum RM-1521.
7. Rosenkrantz, D., et al. 1977. SIAM J. Computing **6**, 563–81.
8. Cayley, A. 1881. Am. J. Math **4**, 266–68.
9. Kruskal, J. 1957. Proc. Am. Math. Soc. **7**, 48–50.
10. Christofides, N. 1976. Report 388, GSIA, Carnegie Mellon University.
11. Lin and Kernighan (1973).
12. Lin and Kernighan (1973).
13. Chandra, B., H. Karloff, C. Tovey. 1994. In: *Proceedings of the Fifth Annual ACM-SIAM Symposium on Discrete Algorithms*. SIAM. 150–59.
14. Helsgaun, K. 2000. Eur. J. Oper. Res. **126**, 106–30.
15. Gates, W. H., C. H. Papadimitriou. 1979. Discrete Math. **27**, 47–57.
16. Kirkpatrick, G., C. D. Gelatt, M. P. Vecchi. 1983. Science **220**, 671–80.
17. Applegate, D., W. Cook, A. Rohe. 2003. INFORMS J. Comput. **15**, 82–92.
18. Hollan, J. 1975. *Adaptation in Natural and Artificial Systems*. University of Michigan Press, Ann Arbor, Michigan.
19. Genetic algorithms for other classes of problems often include a "mutation" step, applied to single members of the population.
20. Nagata, Y. 2006. Lect. Notes Comput. Sci. *4193*, 372–81.
21. See http://www.aco-metaheuristic.org.
22. http://comopt.ifi.uni-heidelberg.de/software/TSPLIB95/.
23. http://www2.research.att.com/~dsj/chtsp/about.html.
24. Johnson, D. S., L. A. McGeogh. 2002. In: G. Gutin, A. Punnen, eds. *The Traveling Salesman Problem and Its Variations*. Kluwer, Boston, MA. 369–443.
25. http://www.tsp.gatech.edu/vlsi/index.html.
26. http://www.tsp.gatech.edu/world/countries.html.

Chapter 5. Linear Programming

1. Grötschel, M. 2006. Notes for a Berlin Mathematical School.
2. Albers and Reid (1986).
3. Dantzig (1991).
4. Carroll, L. 1865. *Alice's Adventures in Wonderland*. Project Gutenberg Edition.
5. The term "nonnegative" is not the same as "positive," which means strictly greater than 0.
6. Safire, W. 1990. "On Language". *New York Times*, February 11.
7. Cottle, R. W. 2006. Math. Program. **105**, 1–8.
8. Dantzig (1991).
9. Gill, P. E., et al. 2008. Discrete Optim. **5**, 151–58.
10. Dantzig (1963).
11. Kolata, G. 1989. *New York Times*, March 12.

12. Williams, H. P. 1999. *Model Building in Mathematical Programming*. John Wiley & Sons, Chichester, UK.

13. Dongara, J., F. Sullivan. 2000. Comp. Sci. Eng. **2**, 22–23.

14. Chvátal, V. 1983. *Linear Programming*. W. H. Freeman and Company, New York.

15. In an LP model we must assume that we can produce and sell fractions of a widget; we will come back to this point at the end of the chapter.

16. http://campuscgi.princeton.edu/~rvdb/JAVA/pivot/simple.html. The tool is a companion to the book: Vanderbei, R. J. 2001. *Linear Programming: Foundations and Extensions*. Kluwer, Boston, MA.

17. Albers and Reid (1986).

18. Dantzig, G. B. 1949. Econometrica **17**, 74–75.

19. Dantzig, G. B. 1982. Oper. Res. Let. **1**, 43–48.

20. Bixby, R. E. 2002. Operations Research **50**, 3–15.

21. Dantzig (1991).

22. The precise statement is that if an LP problem and its dual problem both have allowable solutions, then they both have optimal solutions and the optimal objective values are equal.

23. Jünger, M., W. R. Pulleyblank. 1993. In: S. D. Chatterji et al., eds. *Jahrbuch Überblicke Mathematik*. Vieweg, Brunschweig/Wiesbaden, Germany. 1–24.

24. http://www.informatik.uni-koeln.de/ls_juenger/research/geodual/.

25. Benoit, G., S. Boyd. 2008. Math. Oper. Res. **33**, 921–31.

26. Michel Goemans of MIT has already proven a number of remarkable results concerning solutions to the subtour LP relaxation.

27. Ziegler, G. 1995. *Lectures on Polytopes*. Springer, Berlin, Germany.

28. Christof, T., G. Reinelt. 2001. Int. J. Comput. Geom. Appl. **11**, 423–37.

29. Dantzig, G. B. 1960. Econometrica **28**, 30–44.

30. Dantzig (1963).

31. http://mat.tepper.cmu.edu/blog/.

32. Another great place to keep up on developments in O.R. is Laura McLay's *Punk Rock Operations Research* blog http://punkrockor.wordpress.com/.

Chapter 6. Cutting Planes

1. Hoffman and Wolfe (1985).

2. Dantzig, G., R. Fulkerson, S. Johnson. 1954. Technical Report P-510, RAND Corporation, Santa Monica, California.

3. Note that the only way a tour can make a subtour inequality have value two is for the tour to enter the subset of cities, visit each city in the subset, and then exit the subset. So within the subset of cities the tour looks like a single path. If we have such a path for each of the three blue sets in the figure, then the yellow set must be crossed at least three times. A tour crosses the border of any set an even number of times, so knowing that it crosses at least three times we also

know it crosses at least four times. Altogether, we get three sets crossed two times each and one set crossed four times, so the total number of crossings is at least ten.

4. Note that each of the three red edges in the triangle cross two borders each and are thus counted twice in the total.

5. Dantzig et al. (1954).

6. Dantzig et al. (1954).

7. Chvátal, V. 1973. Math. Program. **5**, 29–40.

8. Grötschel, M., M. W. Padberg. 1979. Math. Program. **16**, 265–80.

9. Grötschel, M., W. R. Pulleyblank. 1986. Math. Oper. Res. **11**, 537–69.

10. To show that an inequality is facet defining we analyze the tours that satisfy the inequality as an equation. Taking the smallest plane that contains these tour points, we check that it lives in a space of dimension exactly one less than the space of the TSP polytope. This is just like the two-dimensional example, where the two corner points on the facet determine a line, that is, a one-dimensional plane.

11. Grötschel, M., M. W. Padberg. 1979. Math. Program. **16**, 265–80.

12. Naddef, D. 2002. In: G. Gutin, A. Punnen, eds. *The Traveling Salesman Problem and Its Variations*. Kluwer, Boston, Massachusetts. 29–116.

13. Schrijver, A. 2002. Math. Program. **91**, 437–45.

14. The subtour inequality for S is identical to the subtour inequality for the set all vertices not in S. We can therefore assume that our sets S contain Phoenix.

15. Flow problems are the cornerstone of a theory of networks that has been steadily improved since the time of the Cold War. Everything you would want to know and more can be found in the text *Network Flows* by R. Ahuja, T. Magnanti, J. Orlin. 1983. Prentice Hall, Englewood Cliffs, New Jersey.

16. Fleischer, L., É. Tardos. 1999. Math. Oper. Res. **24**, 130–48.

17. Letchford, A. L. 2000. Math. Oper. Res. **25**, 443–54.

18. Boyd, S., S. Cockburn, D. Vella. 2007. Math. Program. **110**, 501–19.

19. Cook, W., D. G. Espinoza, M. Goycoolea. 2007. INFORMS J. Comp. **19**, 356–65.

20. Edmonds, J. 1965. Canadian J. Math. **17**, 449–67.

21. Edmonds (1991).

22. Gomory (1966).

23. Khachiyan's algorithm is based on a general technique known as the *ellipsoid method*, developed by David Yudin and Arkadi Nemirovski.

24. Grötschel, M., L. Lovász, A. Schrijver. 1993. *Geometric Algorithms and Combinatorial Optimization*, 2nd edition. Springer, Berlin, Germany.

25. Gomory, R. E. 2010. In: Jünger et al., eds. *50 Years of Integer Programming 1958–2008*. Springer, Berlin. 387–430.

26. The idea of pushing in linear inequalities by rounding down the right-hand side was developed into an exquisite theory by Vašek Chvátal in the early 1970s.

Chapter 7. Branching

1. Little, J. D. C., et al. 1963. Operations Research **11**, 972–89.
2. Eastman, W. L. 1958. *Linear Programming with Pattern Constraints.* Ph.D. thesis. Department of Economics, Harvard University, Cambridge, Massachusetts.
3. Little, J. D. C., et al. 1963. Op. Res. **11**, 972–89.
4. Padberg, M., G. Rinaldi. 1987. Oper. Res. Let. **6**, 1–7.
5. Land, A. H., A. G. Doig. 1960. Econometrica **28**, 497–520.
6. Land, A. H., A. G. Doig. 2010. In: Jünger et al., eds. *50 Years of Integer Programming 1958–2008.* Springer, Berlin. 387–430.

Chapter 8. Big Computing

1. Turing Award Lecture delivered in 1985, available in Karp (1986).
2. Grötschel, M., G. L. Nemhauser. 2008. Discrete Optim. **5**, 168–73
3. Applegate, D., et al. 1995. DIMACS Technical Report 95-05. DIMACS, Rutgers University.
4. Held, M., R. M. Karp. 1971. Math. Program. **1**, 6–25.
5. Karp (1986).
6. Interview with Richard Karp, October 24, 2009.
7. Camerini, P. M., et al. 1975. Math. Program. Study **3**, 26–34.
8. Miliotis, P. 1978. Math. Program. **15**, 177–88.
9. Grötschel, M. 1980. Math. Program. Study **12**, 61–77.
10. Padberg, M. W., S. Hong. 1980. Math. Program. Study **12**, 78–107.
11. Padberg, M. 2007. Ann. Oper. Res. **14**, 147–56.
12. Crowder, H., M. W. Padberg. 1980. Management Science. **26**, 495–509.
13. Grötschel, M., O. Holland. 1991. Math. Program. **51**, 141–202.
14. Padberg, M., G. Rinaldi. 1991. SIAM Review **33**, 60–100.
15. The RSA Factoring Challenge consisted of numbers that the computer-security firm RSA believed to be difficult to factor into their prime components. The firm offered prize money for the successful factorization of each number.
16. Applegate, D. L., et al. 2009. Op. Res. Let. **37**, 11–15.
17. http://www.tsp.gatech.edu/data/art/index.html.

Chapter 9. Complexity

1. Edmonds (1991).
2. Turing, A. M. 1936. Proc. Lond. Math. Soc. **42**, 230–65.
3. Alonzo Church's work was carried out at about the same time as Turing's. Church described computability in a different framework that Turing later showed to be equivalent to his concept of a Turing machine.
4. Beckmann, M., T. C. Koopmans. 1953. Cowles Commission: Econ. No. 2071.

5. Flood (1954).
6. Edmonds (1991).
7. Decision problems do not directly capture the TSP, but we can always find the length of a shortest tour by asking a sequence of yes/no questions.
8. The notion of polynomial-time verifiable problems was also developed independently by Leonid Levin.
9. Cook, S. 1971. In: *Proceedings of the 3rd Annual ACM Symposium on the Theory of Computing.* ACM Press, New York. 151–58.
10. Karp (1972).
11. Garey and Johnson (1979) provides a great introduction to the theory of computational complexity. An excellent recent treatment is Arora and Barak (2009).
12. http://www.win.tue.nl/~gwoegi/P–versus-NP.htm.
13. "Computational Complexity" chapter in Lawler et al. (1985).
14. One way to see this result is to note that the Hamiltonian-circuit problem on a grid graph can be solved as a Euclidean TSP.
15. Held and Karp (1962).
16. A nice reference for this type of work is Woeginger, G. J. 2003. Lect. Notes Comp. Sci. **2570**, 185–207.
17. This statement by Randall Munroe is viewed by moving a mouse over the image of the TSP strip, Number 399, on the xkcd.com Web site.
18. A general Hamiltonian-circuit problem can be encoded by assigning cost 0 to all pairs of cities corresponding to edges in the graph and assigning cost 1 to all other pairs of cities. A Hamiltonian circuit has cost 0, and any non-optimal tour has cost at least 1, so a method that returns a solution within any fixed percentage of optimality will provide an answer to the yes/no question.
19. The exact value is 220/219.
20. Arora, S. 1998. J. ACM **45**, 753–82. See also Mitchell, J. 1999. SIAM J. Computing **28**, 1298–309.
21. Adleman, L. M. 1994. Science **266**, 1021–24.
22. Baumgardner, J. et al. 2009. J. Biol. Eng. **3**, 11.
23. Aono, M., et al. 2009. New Generation Comp. **27**, 129–57.
24. Haist, T., W. Osten. 2007. Optics Express **15**, 10473–82.
25. Oltean, M. 2008. Natural Comp. **7**, 57–70. Oltean's work also appeared in a 2006 conference proceedings, predating the study of Haist and Osten.
26. Aaronson, S. 2008. Sci. Am., March, 62–69.
27. Deutsch, D. 1991. Phys. Rev. D **44**, 3197–217.

Chapter 10. The Human Touch

1. Sheffield, C. 1996. Mood Indigo thoughts on the Deep Blue/Kasparov match. IBM.
2. This work helped to create the *Journal of Problem Solving*, edited by Zygmunt Pizlo, one of the leaders of human TSP research.

3. Vickers, D., et al. 2001. Psychol. Res. **65**, 34–45.
4. van Rooij, I., et al. 2006. J. Prob. Solv. **1**, 44–73.
5. MacGregor, J. N., T. Ormerod. 1996. Percept. Pyschophys. **58**, 527–39.
6. Wiener, J. M., N. N. Ehbauer, H. A. Mallot. 2009. Psychol. Res. **77**, 644–58.
7. Vickers, D., et al. 2004. Pers. Indiv. Differ. **36**, 1059–71.
8. Reitan, R. M., D. Wolfson. 1993. *The Halstead-Reitan Neuropsychological Test Battery: Theory and Clinical Interpretation.* Neuropsychology Press, Tucson, Arizona, USA.
9. Butler, M., et al. 1991. Prof. Psychol.-Res. Pr. **22**, 510–12.
10. Menzel, E. W. 1973. Science **182**, 943–45.
11. Gibson, B. M., et al. 2007. J. Exp. Psychol. Anim. B. **33**, 244–61.

Chapter 11. Aesthetics

1. Nešetřil (1993).
2. *New York Times*, November 17, 1995, page C60.
3. *Proof Positive: Forty Years of Contemporary American Printmaking at ULAE, 1957–1997.* Corcoran Gallery of Art, Washington, D.C., 1997.
4. Paula Cooper Gallery, Julian Lethbridge, March 26–April 24, 1999.
5. Harris, S. 2007. Art in America *95*, issue 10, page 214.
6. *New York Times*, November 17, 1995, page C60; *Baltimore Sun*, March 31, 1996.
7. Bosch, R. 2010. Embrace. *2010 Joint Mathematics Meetings Art Exhibition Catalog.*
8. Kaplan, C., R. Bosch. 2005. Proceedings of the Bridges 2005 Conference.
9. Bosch, R. 2009. Proceedings of the Bridges 2009 Conference.
10. Galanter, P. 2009. Artist text. Artware 5 Exhibition, Lima, Peru, May 2009.
11. Their process is based on a placement algorithm of A. Secord, using weighted Voronoi diagrams.
12. Nešetřil (1993).
13. Korte, B. 1991. *Mathematics, Reality, and Aesthetics – A Picture Set on VLSI-Chip-Design.* Springer Verlag, Berlin.
14. Nešetřil (1993).

Chapter 12. Pushing the Limits

1. June 25, 1956, letter from D. R. Fulkerson to B. Zimmern of Neuilly, France.
2. *New York Times*, March 12, 1991, G. Kolata.
3. Science Blog, June 8, 1998. http://www.scienceblog.com.

Bibliography

[1] Albers, D. J., C. Reid. 1986. An interview with George B. Dantzig: The father of linear programming. The College Mathematics Journal **17**, 293–314.

[2] Applegate, D. L., R. E. Bixby, V. Chvátal, W. Cook. 2006. *The Traveling Salesman Problem: A Computational Study*. Princeton University Press, Princeton, New Jersey.

[3] Arora, S., B. Barak. 2009. *Computational Complexity: A Modern Approach*. Cambridge University Press, New York.

[4] Biggs, N. L., E. K. Lloyd, R. J. Wilson. 1976. *Graph Theory 1736–1936*. Clarendon Press, Oxford, UK.

[5] Chvátal, V. 1973. Edmonds polytopes and a hierarchy of combinatorial problems. Discrete Mathematics **4**, 305–37.

[6] Clay Mathematics Institute. 2000. Millennium problems. http://www.claymath.org/millennium/.

[7] Dantzig, G., R. Fulkerson, S. Johnson. 1954. Solution of a large-scale traveling-salesman problem. Operations Research **2**, 393–410.

[8] Dantzig, G. B. 1963. *Linear Programming and Extensions*. Princeton University Press, Princeton, New Jersey.

[9] Dantzig, G. B. 1991. Linear programming: the story about how it began. J. K. Lenstra et al., eds. *History of Mathematical Programming—A Collection of Personal Reminiscences*. North-Holland. 19–31.

[10] Edmonds, J. 1991. A glimpse of heaven. J. K. Lenstra et al., eds. *History of Mathematical Programming—A Collection of Personal Reminiscences*. North-Holland. 32–54.

[11] Flood, M. 1954. Operations research and logistics. *Proceedings of the First Ordnance Conference on Operations Research*. Office of Ordnance Research, Durham, North Carolina. 3–32.

[12] Garey, M. R., D. S. Johnson. 1979. *Computers and Intractability: A Guide to the Theory of NP-Completeness*. Freeman, San Francisco, California.

[13] Gomory, R. E. 1966. The traveling salesman problem. *Proceedings of the IBM Scientific Computing Symposium on Combinatorial Problems*. IBM, White Plains, New York. 93–121.

[14] Grötschel, M., O. Holland. 1991. Solution of large-scale symmetric travelling salesman problems. Mathematical Programming **51**, 141–202.

[15] Held, M., R. M. Karp. 1962. A dynamic programming approach to sequencing problems. Journal of the Society of Industrial and Applied Mathematics 10, 196–210.

[16] Hoffman, A. J., P. Wolfe. 1985. History. In: Lawler et al. (1985), 1–15.

[17] Karp, R. M. 1972. Reducibility among combinatorial problems. In: R. E. Miller, J. W. Thatcher, eds. *Complexity of Computer Computations*. IBM Research Symposia Series. Plenum Press, New York. 85–103.

[18] Karp, R. M. 1986. Combinatorics, complexity, and randomness. Communications of the ACM 29, 98–109.

[19] Lawler, E. L., J. K. Lenstra, A. H. G. Rinnooy Kan, D. B. Shmoys, eds. 1985. *The Traveling Salesman Problem*. John Wiley & Sons, Chichester, UK.

[20] Lin, S., B. W. Kernighan. 1973. An effective heuristic algorithm for the traveling-salesman problem. Operations Research 21, 498–516.

[21] Mahalanobis, P. C. 1940. A sample survey of the acreage under jute in Bengal. Sankhya, The Indian Journal of Statistics 4, 511–30.

[22] Menger, K. 1931. Bericht über ein mathematisches Kolloquium. Monats-hefte für Mathematik und Physik 38, 17–38.

[23] Nešetřil, J. 1993. Mathematics and art. In: *From the Logical Point of View 2,2*. Philosophical Institute of the Czech Academy of Sciences, Prague.

[24] Reid, C. 1996. *Julia: A Life in Mathematics*. The Mathematical Association of America, Washington, D.C.

[25] Schrijver, A. 2003. *Combinatorial Optimization: Polyhedra and Efficiency*. Springer, Berlin, Germany.

[26] Spears, T. B. 1994. *100 Years on the Road: The Traveling Salesman in American Culture*. Yale University Press, New Haven, Connecticut.

Index